# SpringerBriefs in History of Science and Technology

More information about this series at http://www.springer.com/series/10085

Herbert Capellmann

# The Development
# of Elementary Quantum
# Theory

 Springer

Herbert Capellmann
Institut f. Theoretische Physik
RWTH Aachen University
Aachen
Germany

ISSN 2211-4564          ISSN 2211-4572  (electronic)
SpringerBriefs in History of Science and Technology
ISBN 978-3-319-61883-8        ISBN 978-3-319-61884-5  (eBook)
DOI 10.1007/978-3-319-61884-5

Library of Congress Control Number: 2017949118

Printed on acid-free paper

This Springer imprint is published by Springer Nature
The registered company is Springer International Publishing AG
The registered company address is: Gewerbestrasse 11, 6330 Cham, Switzerland

# Contents

# Abstract

Planck's introduction of the quantum of action in 1900 was followed by 25 years of trial and error in quest of the understanding of the quantum world; different ideas and directions had to be pursued until the path leading to the elementary quantum theory was discovered. Radical changes away from traditional perceptions about natural phenomena were necessary, and the entire system of basic concepts in classical physics had to be abandoned and replaced by a new mode of thought. Continuity and determinism of classical laws were no longer applicable on the quantum scale, where dynamical behavior proceeds by discontinuous and statistical quantum transitions. Albert Einstein laid the essential foundations for the new concept; Max Born made the decisive step further leading to the breakthrough in 1925. The development of the ideas, which eventually resulted in the elementary quantum theory in 1925/26, will be described, relying on original publications and letters written during that period in time by the major contributors. The fundamental laws of quantum theory derived by Max Born and Pascual Jordan may mathematically be represented in many different ways, and particular emphasis is given to the distinction between physical content and mathematical representation.

# Chapter 1
# Introduction and Outline

**Abstract** The scientific development of elementary Quantum Theory will be described, relying on original publications of the major contributors and letters during the same period of time. Recollections from later times are omitted intentionally. Years or decades past have the inevitable tendency to blur and often embellish the glorious past; it is attempted to avoid these effects. The main emphasis will be on physical concepts and ideas; mathematical formulae will be kept to a strict minimum, essential for the understanding. It will be stressed that mathematical representations are interchangeable, the physical content may be described in many different ways.

**Keywords** Planck's radiation law · Planck's quantization hypothesis · Einstein's light quanta · "Old Quantum Theory" · Matrix mechanics · Wave mechanics · "Measurement problem" · Uncertainty relations · Complementarity principle · Copenhagen interpretation · Space-time continuum

Quantum Theory emerged from the endeavor to understand the spectral distribution of radiation in thermal equilibrium with matter. Max Planck's discovery of the radiation law introduced a new universal constant, the quantum of action, marking the start to the "quantum age". The next revolutionary step was taken by Einstein, who recognized that light is constituted of elementary objects, light quanta, having particle properties. During the following years atomic and molecular spectroscopy became the major experimental technique to obtain further insight into the behavior on atomic scales. Nevertheless, the vast literature on the historical development of elementary Quantum Theory—written by practicing scientists, historians, and philosophers of science—typically deals with Quantum Mechanics alone; quantization of radiation is assumed to have played no role.

The central elements of the orthodox accounts may be condensed briefly into the following: The path to a valid Quantum Mechanics in 1925/26 is argued to start from the old Bohr-Sommerfeld Quantum Theory of 1913–1924, which rejected Einstein's light quanta; radiation was believed to retain classical properties. A decisive breakthrough—attributed to Heisenberg's paper of July 1925—is seen to provide the basis for a calculational scheme, matrix mechanics. Schrödinger's wave mechanics, which followed almost instantly, is seen as an independent achievement.

© The Author(s) 2017
H. Capellmann, *The Development of Elementary Quantum Theory*,
SpringerBriefs in History of Science and Technology,
DOI 10.1007/978-3-319-61884-5_1

The recognition of the equivalence of matrix mechanics and wave mechanics then follows. The next milestone is attributed to Heisenberg's uncertainty relations, which are claimed to arise from unavoidable disturbances caused by any measurement of physical variables, leading to the notorious "measurement problem". Bohr's Complementarity principle, attributing dual properties—both wave and particle-like—to photons, electrons, and other particles, completes the so-called Copenhagen interpretation of Quantum Mechanics. The orthodox view is summed up in Max Jammer's book, published in 1966, "The Conceptual Development of Quantum Mechanics":

> "...the conceptual situation as brought about by the establishment of the so-called Copenhagen interpretation ..... is de fact the only existing fully articulated consistent scheme of conceptions that brings into order an otherwise chaotic cluster of facts and makes it comprehensible".

This view will be challenged in the following. It will be shown that Einstein's contributions between 1905 and 1924 provided the basis for Max Born's radically new physical concepts, which relied on Einstein's elementary Quantum Optics to define a path towards Quantum Mechanics. In 1905 Einstein introduced the photon; radiation is quantized, consisting of particles with energies and momenta. In 1917 Einstein's paper "On the Quantum Theory of Radiation" contains the necessary conditions for thermal equilibrium between radiation and matter. Einstein's Quantum Optics provided the starting point for Max Born towards a valid Quantum Theory, combining Quantum Mechanics and Quantum Optics. Born concluded that quantization of action required truly discontinuous behavior of all phenomena in nature; the quantal behavior of matter and radiation are mutually dependent, only their combination may provide a consistent theory. All transitions require the action variables to change by integer values of Planck's quantum of action $h$. Atomic systems can no longer be described by position and time variables varying continuously, but by quantum numbers; discontinuous transitions between different states are describable only by probabilities. The space-time continuum has lost its meaning on elementary scales: In order to define a precise point in space or a precise instant of time, we can only rely on the objects, which nature provides; these objects themselves are subject to quantum laws. A definite point in a space-time continuum looses its meaning on atomic scales, because we need atoms or electrons for their definition; but atoms and electrons themselves do not have precise space-time variables. This understanding led to the Quantum theory of Max Born, Werner Heisenberg, and Pascual Jordan in 1925.

The role of Heisenberg was ambiguous as an intermediate between Göttingen, where he was Born's Assistent, and Copenhagen, where he collaborated with Bohr. Already in 1925 he was primarily influenced by Bohr's thinking and ideas, less by Born's more radical physical concepts. Substantial differences in concepts and understanding remained between Born-Jordan on one side and Bohr-Heisenberg on the other.

This book will proceed as follows:

The scientific development of elementary Quantum Theory will be described, relying on original publications of the major contributors and letters during the same period of time. Recollections from later times are omitted intentionally. Years or decades past have the inevitable tendency to blur and often embellish the glorious past; it is attempted to avoid these effects. The main emphasis will be on physical concepts and ideas; mathematical formulae will be kept to a strict minimum, essential for the understanding. It will be stressed that mathematical representations are interchangeable, the physical content may be described in many different ways.

The very brief Chap. 2 stresses the fundamental differences between continuous and deterministic classical physics to discontinuous and statistical quantum properties.

Chapter 3 starts with a brief account of the path leading to Planck's introduction of the quantum of action. Extensive discussions of Einstein's contributions between 1905 and 1924 follow; they demonstrate the mutual dependence of quantization of radiation and matter. The introduction of the photon in 1905 is extended to include the application of quantization to the properties of matter in 1907. Of decisive importance is Einstein's Quantum Theory of radiation of 1917, which will serve as starting base for the future matrix mechanics. Further milestones are Einstein's application of Bose statistics to particles in 1924, the recognition of quantum theoretical indistinguishability, and the predictions of Bose-Einstein condensation and interference-like phenomena for particles with finite mass.

Chapter 4 provides a brief description of the central elements of the Bohr-Sommerfeld "Old Quantum Theory".

Chapter 5 contains the background and the essential publications for the development of matrix mechanics from 1919 to 1925. The path leads from Born's suggestion in 1919 that the space-time continuum looses its meaning on the quantum scale to the implementation of discontinuous quantum transitions ("*Quantensprünge*") in 1924, when the term "Quantum Mechanics" ("*Quantenmechanik*") is introduced. In June 1925 Born and Jordan introduce discontinuous "quantum vectors" ("*Quantenvektoren*")—which will later become "matrix elements"—to combine Einstein's elementary Quantum Optics with Born's "Quantenmechanik". Heisenberg's "reinterpretation paper" ("*Umdeutung*") of 1925 introduces matrix multiplication rules for the "quantum vectors", which are used by Born and Jordan in September 1925 to achieve the final breakthrough: Commutation relations and quantum equations of motion. The "three men's paper" ("*Dreimännerarbeit*") of Born, Heisenberg, Jordan of November 1925 completes the formal development of matrix mechanics.

Chapter 6 contains continuous representations of the new quantum laws. Kornel Lanczos pointed out this possibility, used by Max Born and Norbert Wiener to represent "time" by a continuous variable again. Schrödinger's wave mechanics—representing position as continuous variable—provided a mathematic method more familiar than the unusual algebra. Schrödinger conceived it as rejection of the radically new concepts; he refused to give up the space-time continuum: "*Das räumlich-zeitliche Denken*" (i.e. the mode of thought relying on continuity in space and time) should remain the only acceptable way to conceive of processes in nature and remain to be the basis for the understanding of the laws of nature.

Chapter 7 discusses the consequences of the basic quantum laws on "wave-like" phenomena and quantum uncertainties. The uncertainty relations are generally attributed to Heisenberg. Heisenberg's justification, however, based on necessary disturbances introduced by the measuring process, will turn out to be erroneous. Bohr's "Complementarity principle", postulating "particle-wave duality", completes the so-called Copenhagen interpretation of Quantum Mechanics. Particle-wave duality will be shown to be a mathematical artifact without physical significance.

Chapter 8 describes the decided opposition of Einstein and Schrödinger to the Copenhagen interpretation. Einstein's understanding of Quantum Theory is described in detail based on his own writings. According to Einstein the **"EPR paper"** (Einstein, Podolsky, Rosen, 1935) was written by Podolsky and does not properly contain Einstein's own stance.

Chapter 9 contains discussions of orthodox accounts of the development of Quantum Mechanics. The classical books by Max Jammer (1966) and B. L. van der Waerden (1967) represent the Copenhagen point of view; the multi-volume work of Jagdish Mehra and Helmut Rechenberg (1982) is similar in general conclusion. The essential differences to the conclusions reached in the present account are illustrated.

Chapter 10 describes later opposition to the Copenhagen interpretation. The logical consistency of the Copenhagen interpretation—in particular Bohr's "philosophical" pronouncements—are viewed with growing skepticism by historians and philosophers of science.

**The appendix** on elementary scattering processes demonstrates the physical content of the fundamental equations of Quantum Theory. The inconsistencies in Heisenberg's and Bohr's arguments become apparent. Disturbance-free measurements are not only possible, but are used routinely in diffraction phenomena, which measure not only average particle positions but their respective quantum uncertainties as well. Particle-wave duality is shown to be neither necessary nor helpful.

# Chapter 2
# The Fundamental Differences Between Classical and Quantum Physics

**Abstract** This very brief chapter stresses the fundamental differences between continuous and deterministic classical physics to discontinuous and statistical quantum properties.

**Keywords** Classical continuity and determinism · Discontinuous and statistical quantum transitions · Probability laws

The basic laws of classical physics relied upon the principle "Natura non facit saltus" (nature does not make jumps); the underlying assumption was the existence of a space-time continuum and all changes in nature should occur continuously within this space-time continuum. Starting towards the end of the 17th century, the classical laws governing these changes were expressed in form of differential equations or variational principles, where infinitesimally small changes of various physical variables are related to each other. Typically these differential equations of classical physics possessed exact solutions for given initial and boundary conditions, at least in principle. This led to the general conclusion that nature is deterministic; the state of nature at any given time was believed to be related in a unique way to its state at any past or future time. Even if the development of statistical thermodynamics related probabilities to thermodynamic variables, these probabilities were meant to describe insufficient knowledge of details due to the large numbers of microscopic particles involved, but deterministic behavior of all individual processes was not questioned.

**Classical physics** relied upon the principles of **continuity and determinism**:

- **Changes of all physical quantities occur continuously in space and time**.
- **The laws determining these changes are deterministic**.

When the microscopic "quantum world" was explored towards the end of the 19th and the beginning of the 20'th century, the basic laws of classical physics (mechanics, electrodynamics and thermodynamics, which had led to great scientific and technological advances during the 19th century) turned out to be unable to describe the observations. The keys, which eventually should lead to the development

© The Author(s) 2017
H. Capellmann, *The Development of Elementary Quantum Theory*,
SpringerBriefs in History of Science and Technology,
DOI 10.1007/978-3-319-61884-5_2

of the elementary Quantum Theory by Max Born, Werner Heisenberg and Pascual Jordan in 1925, are contained in:

The basic principles of **Quantum physics**:

- **On the elementary level all changes in nature are discontinuous, consisting of quantized steps: "quantum transitions".**
- **The occurrence of these quantum transitions is not deterministic, but governed by probability laws.**

Continuity and determinism had to be abandoned, which amounted to a radical change away from all traditional concepts about the laws of nature, and it is not surprising that it took several decades from Planck's quantum hypothesis in 1900 until Born, Heisenberg, and Jordan formulated the elementary Quantum Theory in 1925.

25 years actually are a rather short time, for a totally *"new mode of thought in regard to natural phenomena"* (Max Born) to develop, and the new theory of Born, Heisenberg, and Jordan was received with great skepticism by a large fraction of the "physical establishment", e. g. Max Planck, Albert Einstein (although Einstein himself had laid the most important foundations for the new concept) and many others.

# Chapter 3
# Planck's Quantum Hypothesis and Einstein's Contributions to the Foundations of Quantum Theory

**Abstract** This chapter starts with a brief account of the path leading to Planck's introduction of the quantum of action. Extensive discussions of Einstein's contributions between 1905 and 1924 follow; they demonstrate the mutual dependence of quantization of radiation and matter. The introduction of the photon in 1905 is extended to include the application of quantization to the properties of matter in 1907. Of decisive importance is Einstein's Quantum Theory of radiation of 1917, which will serve as starting base for the future matrix mechanics. Further milestones are Einstein's application of Bose statistics to particles in 1924, the recognition of quantum theoretical indistinguishability, and the predictions of Bose-Einstein condensation and interference-like phenomena for particles with finite mass.

**Keywords** Planck's constant · Boltzmann constant · Planck resonators · Einstein's quanta of light · Photo-electric effect · Wave optics · Maxwell theory · Interference · Specific heat · Momentum and energy fluctuations of the radiation field · Emission and absorption · "Particle-wave duality" · Quantum theory of radiation · Photon momentum · Transition probabilities · Bose statistics · Fermi statistics · Indistinguishability · Bose-Einstein condensation · De Broglie's concept of "phase-waves"

At the meeting of the Deutsche Physikalische Gesellschaft on December 14th 1900 Max Planck presented the derivation of his radiation law, based on a quantum hypothesis and defining a new universal constant, the quantum of action $h$, now called Planck's constant (**Verh. D. Phys. Ges. 2, 553–563, 1901**). Planck's aim had been to explain the spectral distribution of electromagnetic radiation in thermal equilibrium with a surface or a gas of given temperature. The quest for the understanding for this "Normal-Spektrum" will be decisive for the development of quantum theory. In 1859 Kirchhoff had concluded that the basic laws of thermodynamics required the spectral distribution of radiation in equilibrium to be a universal function of frequency and temperature alone. In 1893 Wilhelm Wien formulated his "displacement law" for the energy density of radiation $u$ as function of frequency $\nu$ and temperature $T$: $u(\nu, T) = a\nu^3 f(\nu/T)$ (Berichte der Berliner Akademie of 9 Feb 1893). In 1896 Wien proposed a special form, "Wien's distribution law": $u(\nu, T) = a\nu^3 e^{-c\nu/T}$ (Ann. d. Phys. u. Chem. 58, 662, 1896). Already before December 1900 Max Planck tried to derive

© The Author(s) 2017
H. Capellmann, *The Development of Elementary Quantum Theory*,
SpringerBriefs in History of Science and Technology,
DOI 10.1007/978-3-319-61884-5_3

the special form of the "Normalspektrum" relying only on the phenomenological laws of thermodynamics. First he aimed for the derivation of Wien's special form, the distribution law. When experimental studies showed, that this distribution law gave too low an intensity for low frequencies, Planck extended his phenomenological arguments—by what he himself called an arbitrary assumption—to arrive at (Verh. d. Phys. Ges. 2, 202, 1900)

$$u(\nu, T) = a\nu^3 \frac{1}{e^{b\nu/T} - 1}.$$  (3.1)

Planck was not satisfied with his "arbitrary assumption" and looked for a deeper understanding. He extended his phenomenological considerations and turned to the methods of kinetic theory and Boltzmann's statistical relation between entropy and probabilities to obtain the correct functional form of the radiation law.

$$u(\nu, T) = \frac{8\pi h\nu^3}{c^3} \frac{1}{e^{h\nu/kT} - 1}.$$  (3.2)

$c$ is the velocity of light, $h$ and $k$ were called "universal constants"; fitting the experimental data available at that time, their numerical values were obtained.

The discovery of the radiation law remains to be Planck's lasting and immense merit; in addition to the introduction of the quantum of action $h$, a second universal constant, $k$ (now called Boltzmann's constant), was defined. Nevertheless, Planck's physical assumptions contained in his derivation were incorrect. He did not question the purely classical nature of electromagnetic radiation, in Planck's view completely described by Maxwell's equations. Planck sought the source for non-classical behavior in the interaction process between the electromagnetic field and elementary resonators emitting and absorbing radiation. Planck suggested that emission and absorption processes should be modified, such that the possible resonator energies of frequency $\nu$ were restricted to integer multiples of $h\nu$. The derivation of the average thermal resonator energy was performed using discrete values of $n h\nu$ (n integer) only. The result obtained was inserted into the equilibrium relation between thermal averages of resonator energy and radiation energy, derived previously from purely phenomenological considerations, which were assumed to remain valid. The correct result for the radiation law obtained by Planck was due to the use of quantized resonator energies, whereas the radiation field was still assumed to be classical.

## 3.1 Einstein's Quanta of Light

It was Albert Einstein, who recognized that the electromagnetic field itself is quantized. On 17 March 1905 he submitted the paper "Über einen die Erzeugung und Verwandlung des Lichtes betreffenden heuristischen Gesichtspunkt" (On a heuristic point of view concerning the creation and conversion of light), published in **Ann.**

**Phys. 17, 132–148, 1905.** Einstein (re-)introduced the particle concept of radiation, claiming that—on the elementary level—light consists of quanta having particle properties. Einstein concluded that the radiation energy consists of

> "... einer endlichen Zahl von in Raumpunkten lokalisierten Energiequanten, welche sich bewegen, ohne sich zu teilen und nur als Ganze absorbiert und erzeugt werden können" (a finite number of energy quanta which are located in points of space, which move without splitting up and which can only be absorbed and created as a whole).

Einstein based his arguments on the experimental information available about the interaction of electromagnetic radiation with matter, in particular experiments on the creation of cathode rays by light (the "photo-electric effect"), the inverse process of cathode luminescence, the ionization of gases by ultraviolet radiation, and photoluminescence. In all these processes energy is exchanged between radiation and point like **particles**. The particle energies in these processes do not depend on the intensity of radiation but on its frequency $\nu$, whereas only the number of particles involved depends on the intensity. Einstein drew the conclusion that the radiation energy cannot be distributed continuously, but is distributed discontinuously in space in units of $\epsilon = h\nu$, in the same way as matter is made up of discrete particles.

This "heuristic point of view" is confirmed by a theoretical analysis of the thermo-dynamic properties of radiation. Einstein used a particular model which contained three different systems: An ideal gas, charged particles oscillating linearly, and the radiation field. Thermal equilibrium between ideal gas and oscillators required that the average kinetic energy of oscillating particle and gas molecule be identical. This condition required—according to classical laws—that the average oscillator energy $\bar{\epsilon}_\nu$ be equal to $kT$. Planck's equilibrium conditions between average energy density of radiation and average resonator energy ($u(\nu, T) = 8\pi\nu^2/c^3 \;\; \bar{\epsilon}_\nu(T)$) was used to obtain the classical result for the expected radiation energy. The "classical limit" of Planck's radiation law for very high temperature and small values of $h\nu/kT$ is shown to be compatible with classical laws and independent of Planck's constant $h$. The radiation intensity becomes very high in this region of high enough temperature or sufficiently low frequency. In the high frequency and low temperature region—i.e. for $h\nu/kT$ large, where Planck's radiation law is equivalent to Wien's—the classical laws fail completely, however. They predict a diverging radiation density, in total con-trast to observations. Instead of diverging, the observed radiation density in thermal equilibrium vanishes exponentially, in accord with Planck's law.

Einstein—in contrast to Planck—identifies the correct reason for the failure of classical theory; he conducts a thermodynamic analysis of radiation for the high frequency region, where the observed radiation density becomes very small. The analysis confirms the existence of light quanta having particle character; the thermo-dynamic properties of radiation are shown to be consistent with the behavior of an ideal gas of particles with energies $h\nu$.

Einstein was well aware that, introducing the particle concept of light quanta, he not only solved the problem of the interaction process of light with matter, he also created a new problem: If light consists of particles, how can wave optics be reconciled with the particle picture? Einstein himself pointed out that his particle concept of light left the question open why wave optics gave an accurate description of many macroscopic phenomena. Einstein stated explicitly, that

"...die Undulationtheorie des Lichtes sich zur Darstellung der rein optischen Phänomene vortrefflich bewährt hat und wohl nie durch eine andere Theorie ersetzt werden wird" (to describe purely optical phenomena the undulation theory of light has proven its worth excellently and most probably will never be replaced by another theory).

Remarkably, Einstein points already into the correct direction to solve this new problem, suggesting that wave optics should apply to averages only.

During the years before 1905, Einstein had made fundamental contributions to a similar problem of thermodynamics (A. Einstein, Ann. d. Phys. 9, 417–433, 1902; 11, 170–187, 1903; 14, 354–362, 1904): How to connect the individual properties of the constituents of matter to the macroscopic laws of phenomenological thermodynamics. Atoms in a gas have energy and momentum; temperature, entropy, pressure are properties of macroscopic averages. Gibbs and Einstein established the connection, based on Boltzmann's relation between probabilities and entropy. Einstein recognized that the equivalent problem had to be solved to reconcile particle properties and wave phenomena of radiation. But the solution to "Einstein's new problem", the connection between the microscopic particle behavior of light and the phenomenology of wave optics, will have to wait until the advent of a valid quantum theory several decades later.

Planck, as well as the vast majority of his contemporaries up to and partly beyond 1925, rejected Einstein's revolutionary concept. From today's perspective it might seem highly peculiar, that Einstein's hypothesis was met with general disbelief for several decades, but we should not underestimate the highly revolutionary character of Einstein's proposal, which, for Max Planck and most of his contemporaries, seemingly negated one of the most important scientific advances of the 19'th century, Maxwell's theory of electromagnetism and the explanation of all wave-optical phenomena, such as diffraction, dispersion, refraction, and reflection.

Let us examine the apparent contradictions between Einstein's hypothesis and the traditional interpretation of radiation in more detail, in particular the central problem of the spectral distribution of radiation in thermal equilibrium with matter. The experimental determination of the spectrum directly assigned wavelengths and frequencies—i. e. wave characteristics—to radiation; these experiments seemingly directly disproved Einstein's hypothesis. Einstein claimed that radiation in thermal equilibrium consists of statistically emitted and absorbed particles, that these particles are noninteracting and having thermal properties of an ideal gas! It seemed inconceivable that noninteracting particles, statistically emitted and absorbed, could provide frequencies and wavelengths; even the basic concept of constructive and destructive interference seemed to be in total contradiction with Einstein's picture.

Einstein conceded, that he did no longer understand the origin of wave optics; but he also concluded that the experimental information about the interaction between radiation and matter was incompatible with wave behavior on elementary scales. This belief was strengthened in the following years, when Einstein made further decisive contributions for the development of quantum theory.

## 3.2 Einstein's Application of Quantization Concepts to the Properties of Matter

In 1907 (Ann. Phys. 22, 1907, 180–190; Ann. Phys. 22, 1907, 800) Einstein applied the quantization concept to oscillations of atoms in solids, necessary for the understanding of thermal properties, in particular the temperature dependence of the specific heat.

When Einstein introduced quanta of light in 1905, Planck's classical equilibrium conditions between average energy density of radiation $u(\nu, T)$ and average resonator energy $(u(\nu, T) = 8\pi\nu^2/c^3 \; \bar{\epsilon}_\nu(T))$ had been used. The classical limit for high temperatures and low frequencies proved to be compatible with classical laws. For high frequencies and low temperatures, however, classical laws failed to properly account for the exponentially small radiation density. Einstein solved this problem by the postulate of quanta of light. But there was another unsolved problem remaining: The classical result for the average oscillator energy $(\bar{\epsilon}_\nu = kT)$ required the contribution to the specific heat to be temperature independent. This conclusion is not only incompatible with measured specific heats of solids, but also with Nernst's third law of thermodynamics. The entropy should vanish for zero temperature, which requires that the specific heat must vanish at least linearly or faster with temperature tending towards zero.

Atoms in solids may perform vibrations around their equilibrium positions; according to the classical laws, a solid containing $N$ atoms should contribute a vibrational specific heat of $N \cdot 3k$ (the Dulong-Petit law), independent of the frequencies of vibration and of temperature. For high enough temperature this classical limit was indeed observed for many solids, but lower temperatures showed more or less strong deviations from classical predictions. In particular experiments on solids containing light atoms—where high frequencies could be expected—showed much lower specific heat for low temperatures.

Einstein drew the conclusion that matter, too, is showing quantal behavior. In order to obtain a qualitative understanding, he replaced the possible oscillation frequencies by one single frequency $\nu_E$ (what he himself called a crude assumption). Planck's quantization hypothesis for oscillators is used to calculate average energies and specific heats. Exponentially small values are obtained for low temperature, i.e. large values of $h\nu_E/kT$, whereas for high enough temperature ($h\nu_E/kT$ small) the classical Dulong-Petit value is reached. The measured specific heat of diamond between 200 and 1300 K is compared to the theoretical results and good agreement is obtained.

Thereby Einstein extended the quantization concept to include all phenomena in physics, radiation, matter, and their mutual interaction. In the first paper cited above, he connected the oscillation processes in matter to emission or absorption of light quanta; the second very short paper added a correction; non-radiative oscillations are equally possible.

## 3.3  The Momentum and Energy Fluctuations of the Radiation Field

**In 1909 (Phys. ZS. 10, 185–193, 1909; and Phys. ZS. 10, 817–826, 1909)** Einstein discussed the energy and momentum fluctuations of the electromagnetic field and showed that the experimentally established radiation law was fully compatible with the particle concept of light quanta, carrying energy $\epsilon = h\nu$ and momentum $\epsilon/c$. It is in this publication that Einstein mentions the momentum of the light quantum explicitly (which he had not done in 1905, although—for Einstein in particular—it was certainly obvious that point-like energy quanta moving with the speed of light must possess momenta compatible with relativity theory). Einstein derived the energy and momentum fluctuations of radiation in equilibrium and connected the fluctuation characteristics to the functional form of the radiation spectrum. He showed that the mean square fluctuations of the energy density $(\delta u)^2$ consisted of the sum of two terms, a first term proportional to the energy $u$, and a second term proportional to $u^2$. The equivalent result was obtained for the pressure fluctuations. In classical wave theory only terms proportional to $u^2$ may arise, fluctuations resulting from interference between waves of comparable wavelengths, i.e. second order effects.

Einstein showed that the linear term proportional to $u$, which dominates for visible light and higher frequencies, is a typical result for independent particles, giving support to the existence of light quanta. He insists that, since the fluctuation results were derived from generally valid fluctuation properties and furthermore only rely on the functional form of Planck's distribution law, classical wave theory is incompatible with the experimentally established from of the spectral distribution. Only for very high radiation intensities (in the classical limit) the term proportional to $u^2$ becomes important. Einstein mentions that, due to the additive connection of the two fluctuation terms, they behave like fluctuations of independent origin.[1]

The second publication mentioned above (Phys. ZS. 10, 817–826, 1909) is a reproduction of a lecture Einstein had given at the annual meeting of the Association of German Scientists (81. Versammlung deutscher Naturforscher und Ärzte, Salzburg, 21 Sept 1909) and concludes with a discussion, where Planck expresses his objections to Einstein's light quanta. Planck confirms that—in his opinion—Maxwell's equations remain to be a complete description of electromagnetic radiation; he concluded that interference phenomena could not possibly result from a theory based of radiation quanta with particle character. Quantum effects should be restricted to the interaction of matter and radiation, in particular the emission and absorption processes. Planck conceded that the interaction between free radiation and matter was still poorly understood; emission and absorption processes should require strong acceleration of "resonators" which should fall outside the traditional description. Planck suggested that these poorly understood processes might be responsible for restricting the resonator energies to integer multiples of elementary quanta.

---

[1] Einstein's fluctuation formulae later will be interpreted to give support for the concept of "particle-wave duality" of photons and other particles. Einstein himself, however, will reject this concept. A detailed discussion will be contained in later chapters.

Einstein's response to Planck indicated how the connection between radiation quanta and Maxwell's equations might be accomplished. Already in his 1905 paper, when Einstein introduced the photon concept, he had explicitly stated that the undulation theory should remain valid for **averages**. Einstein suggested that electromagnetic waves should be composed of a macroscopic number of quanta, which—on average—form vector fields; the effective equations for the resulting fields should coincide with Maxwell's equations. He compared this with the origin of electrostatic fields produced by charged macroscopic bodies, which themselves consist of a macroscopically large number of elementary particles carrying finite charges. The fields describe the averages. Maxwell's equations thereby are considered to play the role of phenomenological equations, independent of the detailed structures on atomic scales. Concerning the interference-like term proportional to $u^2$ in the fluctuation spectra, Einstein remarked that he saw no contradiction to the particle concept of light quanta.

## 3.4 The Quantum Theory of Radiation

Einstein's papers "Zur Quantentheorie der Strahlung" (Verh. d. D.Phys. Ges. 18, 318–323, 1917; and Phys. Z. 18, 121–128, 1917) will be of decisive importance for the Quantum Theory to be developed by Max Born, Werner Heisenberg, and Pascual Jordan. These publications established the conditions necessary for thermal equilibrium between radiation and matter. Energy and momentum exchange between radiation and matter is described by emission and absorption of photons; conservation of energy and momentum for all individual processes are required to assure that thermal equilibrium is maintained.

The first paper addresses energy exchange between matter and radiation; it contains the first derivation of the complete form of Planck's radiation law, which is based on physical arguments fully consistent with the future development of Quantum Theory. The essential assumptions are the following:

(a) Energy conservation is valid for all transitions between quantum states emitting or absorbing a photon.
(b) Spontaneous transitions from a quantum state of energy $\epsilon_m$ must be to a state of lower energy $\epsilon_n$, the corresponding probability is $A_m^n$.
(c) Transitions with energy transfer $(\epsilon_m - \epsilon_n)$ induced by the radiation field $u$ may occur with probability $B_n^m \cdot u$ for absorption and probability $B_m^n \cdot u$ for emission.
(d) The relative probability for a particular quantum state of energy $\epsilon_n$ is given by $exp(-\epsilon_n/kT)$.

Einstein shows that the induced emission and absorption probabilities have to fulfill the condition $B_n^m = B_m^n$. Requiring thermal equilibrium leads to

$$u = \frac{A_m^n}{B_n^m} \frac{1}{e^{\frac{\epsilon_m - \epsilon_n}{kT}} - 1} \tag{3.3}$$

Remark that there are no special assumptions made about the specific properties of the various quantum states. Einstein's derivation has general validity for all types of spectra, all possible energy transfers between matter and radiation field, and all temperatures.

Further conclusions are obtained from the limiting cases of low temperatures, where Wien's radiation law is valid, and the classical limit of high temperatures. Wien's radiation law for low temperature and high frequencies requires that the energy transfer $(\epsilon_m - \epsilon_n)$ is equal to $h\nu$ and that $A_m^n / B_m^n$ must have the form $a\nu^3$, where $a$ and $h$ have to be universal constants. In the limit of high temperature and low frequencies the classical limit must be recovered, which determines the factor $a$; finally Planck's radiation law valid for all frequencies and temperatures is recovered.

The second paper cited above (Phys. Z. 18, 121–128, 1917) deals with momentum exchange between matter and radiation. It confirms the existence of the photon momentum; momentum conservation for all individual transitions is shown to be necessary for the establishment of thermal equilibrium between radiation and a gas.

**Transition probabilities** which will play a crucial role in the development of Quantum Theory in 1925. But it is important to notice the distinction between Einstein's use of "probabilities" and the meaning assigned to transition probabilities in the future Quantum Theory. In Einstein's view, probabilities reflect thermodynamic arguments. If we want to connect phenomenological laws of thermodynamics with the underlying microscopic processes, it is typically sufficient to restrict the microscopic description to probabilities, the phenomenological laws on macroscopic scales being independent of a detailed knowledge of all microscopic processes involved. Einstein maintained that the microscopic behavior should be governed by deterministic laws in principle; concerning his own theory he conceded that "the theory has the following weaknesses: First, it does not provide a connection to wave theory; second, it leaves time and direction of the elementary processes to chance".

Later, Max Born will make the decisive step further; he will conclude that discontinuous and statistical behavior is a fundamental property of nature for all elementary processes, classical laws being valid approximately for macroscopic averages only. This principle will be essential for the Quantum Theory of Born, Heisenberg, and Jordan developed in 1925, which will assign purely statistical laws to all individual quantum transition. Einstein, however, will refuse to take this step.

## 3.5  Bose Statistics, Indistinguishability of Particles, and Bose-Einstein Condensation

Einstein's conviction concerning the particle character of radiation quanta even led him to predictions about totally new quantum phenomena, which were to be confirmed only much later. In 1924 Satyendra Nath Bose sent a manuscript to Einstein, which opened the new field of quantum statistics, in particular Bose statistics. Einstein recognized the importance of Bose's discovery, he translated the paper himself[2] and had it published (**S. N. Bose, "Plancks Gesetz und Lichtquantenhypothese", Z. Phys. 26, 178–181, 1924**). Bose's publication contained the first direct derivation of Planck's radiation law, which connected exclusively particle properties of photons directly to the thermodynamic properties of radiation,[3] at the expense of requiring new quantum statistical principles:

(a) Phase space (e.g. spanned by position and momentum variables) is divided into unit volumes $h^3$; and

(b) physical states are characterized by the occupation numbers of the different unit volumes of size $h^3$.

The first principle might be called "phase-space quantization"; a particle is no longer characterized by **precise** values of position and momentum, the uncertainty within the finite volume $h^3$ represents a **quantum uncertainty**. The second principle implies **indistinguishability** of identical particles. These principles will turn out to be of general validity; "Bose-statistics" allows multiple occupation numbers. Later, "Fermi statistics" (E. Fermi, Z. Phys. 36, 902, 1926, rec 24 Mar 1926) will allow single occupancy only.

Two publications by Einstein followed quickly (**Sitz. Berlin Ak. d. Wiss. 10-07-1924, and 08-01-1925**), which applied Bose statistics to the Quantum theory of ideal gases. The first paper discusses the Bose gas in its "normal state", at temperatures above a critical temperature $T_c$; the second paper treats lower temperatures as well and predicts the macroscopic quantum phenomenon of **Bose-Einstein condensation**. It should take 70 more years until Bose-Einstein condensation could be realized experimentally.

Bose statistics requires the average number of atoms with kinetic energy $\epsilon$ to be

$$n(\epsilon) = \frac{1}{e^{\frac{\epsilon-\mu}{kT}} - 1},\tag{3.4}$$

where $\mu$ is the chemical potential, which controls the average density. For photons the chemical potential vanishes; photons can be absorbed and emitted and the photon density is determined by the requirement of thermal equilibrium. The average density of the ideal atomic gas may be controlled by experimental conditions; the chemical potential is volume and temperature dependent and has to fulfill the condition $\mu \leq 0$. Einstein introduces a "degeneracy parameter" $\lambda$, defined by $\lambda = exp(\mu/kT)$, such that $0 \leq \lambda \leq 1$. The first paper discusses the "normal state", where $\lambda \neq 1$. The thermodynamic properties are derived and compared with classical theory. For all values of $\lambda$, lower energy states are preferred, compared to the classical Boltzmann distribution. The classical limit is obtained, if the $-1$ in the denominator of Eq. (**4**) is neglected, and if the degeneracy parameter $\lambda$ is neglected as well.

---

[2]The paper finishes with: "translated by A. Einstein."

[3]Einstein's own derivation of 1916 still had to use experimental information from Wien's displacement law.

Bose's paper and Einstein's first paper had been criticized by Ehrenfest and others for their use of Bose statistics, which treats photons and atoms no longer as statistically independent objects. Einstein's insists; the introduction of the second paper states that Bose's theory, treating photons as a quantum gas, should find its complete analogy in gases of atoms and molecules. He explicitly requires, that photons or other identical particles are not statistically independent, which will lead him to the prediction of Bose-Einstein condensation and diffraction phenomena with particles of finite mass.

Einstein compares the properties of the quantum gas to those of the classical theory. Entropy calculated by classical statistics is not an extensive quantity; to obtain proper thermodynamic behavior, classical statistics has to eliminate the statistical independence of identical particles artificially. Quantum statistical indistinguishability eliminates this problem automatically, for Einstein a decisive argument. Phenomenological thermodynamics and its connection to microscopic properties via Boltzmann's relation between probabilities and entropy had be be fulfilled. The solution to a second thermodynamic requirement encouraged Einstein to predict the macroscopic quantum phenomenon of Bose-Einstein condensation. Nernst's third law of thermodynamics requires the entropy to vanish for temperature tending towards zero. Einstein's theory predicted that lowering temperature—keeping the volume fixed— a critical temperature $T_c$ will be reached. Below $T_c$, only a fraction of the atoms will have finite kinetic energy, whereas the rest is condensed into the lowest energy state. Einstein compares this phenomenon to classical condensation of a saturated gas into the liquid phase. At zero temperature all particles of the ideal quantum gas will be condensed into the lowest energy state available, fulfilling the requirement of vanishing entropy.

The solution to a third problem concerned Einstein's own theory of light quanta. In 1909 he had calculated the fluctuation characteristics of energy and momentum, imposed by the functional form of Planck's radiation law. At low density, where Wien's law is applicable, a term proportional to the density $n$ dominates, characteristic for an ideal gas of particles. But there existed a second term, not negligible at high density, proportional to $n^2$, indicating a mutual influence of the constituents. Such a term is the only one classical radiation theory may produce, resulting from interference of waves of comparable wavelengths. Now Einstein calculated the fluctuation spectrum of the ideal quantum gas; not surprisingly, again two additive terms were obtained. The first term was proportional to the particle density $n$, similar to the results of classical theory of noninteracting particles. But again, there existed another term of second order in particle density $n^2$, indicating some type of mutual influence between particles. Einstein showed that indistinguishability introduced by Bose-statistics produces correlation effects similar to an effective attraction. Noninteracting Bose-particles have the tendency to prefer each other's company; the probability to find a second particle close to a first particle is enhanced, compared to classical statistics.[4] This phenomenon explains the appearance of the second term

---

[4]This effect is now called "photon bunching"; direct observation was obtained by Robert Hanbury Brown and Richard Twiss in 1956.

proportional to $n^2$ in the fluctuation characteristics. The new quantum statistical principle is identified as the cause, valid for particles with finte mass as well as for photons. Wave characteristics are not required.

The success of the quantum statistical principle encouraged Einstein to make a further prediction. In 1905 and in 1909, he had argued that Maxwell Theory should remain to be valid as phenomenological theory for averages over large number of photons, even though he was unable to provide the theoretical connection. Now Einstein extended the particle analogy between photons and particles with finite mass; he suggested that interference-like effects similar to wave optics should be expected for particles with finite mass as well. In this context Einstein refers to the thesis of Louis de Broglie, of which Einstein had obtained a copy prior to publication. De Broglie's concept of "phase-waves" will be discussed later in the chapter about wave mechanics.

# Chapter 4
# The "Old Quantum Theory"

**Abstract** This chapter provides a brief description of the central elements of the Bohr-Sommerfeld "Old Quantum Theory".

**Keywords** Ritz' combination principle · Bohr's atomic model · Bohr's frequency condition · Electron orbits · "Virtual oscillators" · Stationary states · Compton effect · Bohr-Kramers-Slater Theory ("BKS-Theory") · "Virtual radiation"

Planck's concept, i.e. classical radiation and continuous electronic oscillations emitting and absorbing classical radiation in some mysterious and unexplained way, formed the basis for the old quantum theory, which dominated quantum theoretical attempts from 1913 up to the year 1925. The attention shifted away from Planck's radiation law to the analysis of spectroscopic data. In 1908 Walter Ritz (Phys. Zeitschr. 9, 521, 1908) discovered that the multitude of observed spectral frequencies characteristic of atoms and molecules could be classified by a combination principle: $v_{ij} = X_i - X_j$. The Balmer-Rydberg series of hydrogen ($v_{nm} = X(1/n^2 - 1/m^2)$, n and m integers) was a special example. Although Einstein's light quanta allowed a simple explanation, this was rejected; Ritz considered radiation to be classical.

In 1913 Niels Bohr proposed an atomic model (**Phil. Mag. 16, 1–25, 1913**), which was based on a physical picture similar to the planetary system, electrons taking the place of planets and the nucleus replacing the sun. Electronic orbits were determined by classical equations of motion. The free radiation field was assumed to be classical, described by Maxwell equations. Bohr rejected Einstein's light quanta and the quantum nature of the electromagnetic field; repeatedly Bohr insists on the **continuous** character of the field, which—in his mind—was the only possibility to explain wave optical phenomena. Like Planck, Bohr considered the interaction process between electrons and radiation to be responsible for quantum effects, not compatible with classical physics.

© The Author(s) 2017
H. Capellmann, *The Development of Elementary Quantum Theory*,
SpringerBriefs in History of Science and Technology,
DOI 10.1007/978-3-319-61884-5_4

Quantum conditions were postulated for the selection of stationary states, i.e. stable atomic configurations, and for the emitted radiation frequencies. Within stationary states electrons were assumed to perform orbits of circular or elliptical shape with rotational frequencies $\omega_n$. No radiation was to be emitted within these stationary states; emission of homogeneous radiation should result from transition processes between two stationary states.

Multiplication of Ritz' combination scheme by Planck's constant resulted in Bohr's quantum condition for the emitted radiation frequencies

$$h\nu_{ij} = W_i - W_j. \tag{4.1}$$

Bohr identified the right hand side as the difference in energy of two stationary states.

We stress the difference in interpretation of this equation, which is used by both Einstein and Bohr, but has different physical significance according to Einstein and Bohr. They agree on the right hand side, which is the energy difference between two states, but they have totally different interpretations for the left hand side. For Einstein $h\nu_{ij}$ is the energy of the photon, i.e. a particle, emitted or absorbed by the transition process; the equation expresses energy conservation. As mentioned before, Bohr was convinced that radiation is classical; only a wave picture should be able to account for the observation of interference effects. Since Bohr's calculations will produce rotational frequencies $\omega_n$ without any relation to the observed radiation frequencies, the condition $h\nu_{ij} = W_i - W_j$ constituted an additional assumption with no apparent physical explanation. Bohr "solved" this problem suggesting "virtual oscillators" responsible for emission and absorption of the observed frequencies.

The first aim was to give an explanation for the Balmer-Rydberg series of hydrogen ($\nu_{nm} = X(1/n^2 - 1/m^2)$, n and m integers); the frequency condition (Eq. 3) required the $W_n$ to be given by $W_n = hX/n^2$. To obtain the desired result, Bohr made the following assumption:

(i) The electron is circling the nucleus with rotational frequency $\omega_n$; classical mechanics relates $\omega_n$ to the energy $W_n$.

(ii) The selection of stationary states requires the $W_n$ to be equal to $n$ times the energy quanta of "Planck oscillators"[1] of frequencies $f_n$, i.e. $W_n = nhf_n$.

(iii) To complete the determination of the $\omega_n$ and $W_n$, a relation between oscillator frequencies $f_n$ and rotational frequencies $\omega_n$ was required. The relation $f_n = \omega_n/2$ resulted in

$$W_n = \frac{2\pi^2 me^4}{h^2} \frac{1}{n^2}. \tag{4.2}$$

---

[1] The "Planck oscillators" are different from the "virtual oscillators"; the latter are supposed to have frequencies equal to radiation frequencies $\nu_{nm}$. The $f_n$ are unrelated to the $\nu_{nm}$.

Inserting these values into the frequency condition ($h\nu_{nm} = W_n - W_m$) reproduced the experimentally observed frequencies of hydrogen. Classical equations of motion were also used to calculate radii and angular momenta of stationary states. The results for electronic radii agreed with experimental estimates of atomic radii. The angular momentum of state $n$ was obtained as $J_n = n\frac{h}{2\pi}$.

Bohr's calculations were partly "successful" due to dimensional reasons. If an energy scale is to be obtained from the available natural constants (electron mass, elementary charge, and Planck's constant), the combination $me^4/h^2$ follows. Integer numbers were obtained invoking energy quanta of Planck oscillators; the factor $1/2$ between $f_n$ and $\omega_n$ was a free parameter and could be chosen to obtain the correct energies, although all physical assumptions were incorrect.

While the heuristic assumptions at first sight seemed to be rather arbitrary, the "success" to reproduce the observed spectral lines for atomic hydrogen was taken as a first step towards the solution of the quantum mystery; similar attempts followed to reproduce the observed spectral lines of other elements. Additional assumptions concerning the nature of multi-electron orbits were added; electrons were positioned in symmetric arrangements (all of them classically unstable, however) on one or more rings or ellipses rotating rigidly.

The theoretical activities in quantum theory for the next 12 years were dominated by extensions of Bohr's concept. Sommerfeld (Ann. Phys. 51, 1–94, 1916) proposed a reformulation of the quantum condition: 'The integral of position times momentum over one period is required to be equal to an integer multiple of $h$', which replaced Bohr's quantization condition and enabled further applications. The detailed analysis of spectroscopic data was seen as the central problem in quantum physics.

The review by Bohr (**Z.Phys.13,117–165, 1923**) restates the basic assumptions of the old quantum theory:

- Classical continuity in space and time is retained for all processes.
- The electromagnetic field is purely classical.
- Periodic orbits of electrons around nuclei are determined by classical equations of motion.
- Stationary states are selected by quantum conditions: The integral of position times momentum over one period is required to be equal to an integer multiple of $h$.
- No radiation is emitted within stationary states.
- "Virtual oscillators"—with frequencies unrelated to the periodicity of the orbits—emit or absorb homogeneous radiation during transitions between stationary states.
- The frequencies of the virtual oscillators and the radiation frequencies are equal to the energy differences between stationary states divided by $h$.
- The transition process itself is continuous having finite duration.

The discovery of the Compton Effect (A. H. Compton, Bull. Nat. Res. Council 4, Nr 20, Oct 1922) and its theoretical interpretation based on Einstein's photon concept (H. A. Compton, Phys. Rev. 21, 483, 1923, and P. Debye, Phys. Z. 24, 161, 1923) generated some support for Einstein (e.g. W. Duane[2], Proc. Natl. Ac. Sc, 9, 158, 1923; W. Pauli, Z. Phys. 18, 272, 1923, and Z. Phys. 22, 261, 1924), but the majority—and in particular Niels Bohr and his followers—remained unconvinced.

The supposedly purely classical nature of the electromagnetic field is also the stumbling block for N. Bohr, H. Kramers, J. C. Slater (the "BKS theory"; Phil. Mag. 47, 785–802, 1924; and Z. Phys. 24, 69–87, 1924), when they try to address the interaction process between radiation and matter in more detail. To justify their assumptions about the continuous and classical character of radiation on one side and the fast transition process between stationary states on the other, additional assumptions are introduced, which—from today's perspective—seem rather bizarre. As before, two different time scales are used, continuity in time is retained for both scales: a "classical" time scale for classical radiation and a "fast" time scale for the finite duration of the transition process between stationary states, associated with "virtual oscillators". Emission of continuous radiation at the oscillation frequency is postulated to be a collective phenomenon involving many atoms simultaneously. An additional radiation field is introduced ("virtual radiation") instantly connecting different atoms and coordinating transitions in distant atoms such that the continuous and classical behavior of electromagnetic radiation is maintained. Energy and momentum conservation are claimed to be absent for individual transitions and are postulated to be valid on average only.

The BKS theory marks the impasse in which the old quantum theory had ended up. The hypothetical electron dynamics in atoms was without any experimental evidence; radiation was still assumed to be classical; classical equations of motion were supplemented by heuristic and unrelated hypotheses; a path towards a general and internally consistent quantum theory seemed not in sight. A fundamentally new concept was necessary to attain the understanding of the quantum world.

---

[2] William Duane was the first to point out that Einstein's particle concept of photons is able to describe the so called "wave optical" observations, which had been interpreted to result from "interference phenomena". Duane's paper will be described later in more detail.

# Chapter 5
# The Quantum Theory of Born, Heisenberg, and Jordan

**Abstract** This chapter contains the background and the essential publications for the development of matrix mechanics from 1919 to 1925. The path leads from Born's suggestion in 1919 that the space-time continuum looses its meaning on the quantum scale to the implementation of discontinuous quantum transitions ("*Quantensprünge*") in 1924, when the term "Quantum Mechanics" ("*Quantenmechanik*") is introduced. In June 1925 Born and Jordan introduce discontinuous "quantum vectors" ("*Quantenvektoren*")—which will later become "matrix elements"—to combine Einstein's elementary Quantum Optics with Born's "Quantenmechanik". Heisenberg's "reinterpretation paper" ("*Umdeutung*") of 1925 introduces matrix multiplication rules for the "quantum vectors", which are used by Born and Jordan in September 1925 to achieve the final breakthrough: Commutation relations and quantum equations of motion. The "three men's paper" ("*Dreimännerarbeit*") of Born, Heisenberg, Jordan of November 1925 completes the formal development of matrix mechanics.

**Keywords** "Quantensprünge", quantum leaps · Elimination of space-time continuum · Matrix mechanics · Quantum dynamics · Hamilton Jacobi formalism · Born's quantization condition · "Quantenoptik" · "Time" and "transition probability per unit time" · "Virtual oscillators" · Final breakthrough · Commutation relations · Quantum equations of motion · Quantization of the radiation field · Eigenvalue theory of Hermitian forms · Selection rules · Statistical treatment of black body radiation · Born-Jordan versus Bohr-Heisenberg

© The Author(s) 2017
H. Capellmann, *The Development of Elementary Quantum Theory*,
SpringerBriefs in History of Science and Technology,
DOI 10.1007/978-3-319-61884-5_5

## 5.1   Born's Discontinuous "Quantenmechanik"

The basic ideas about purely discontinuous and statistical behavior of all elementary processes in nature are due to Max Born. Already several years before 1925 Born questioned the applicability of the classical concepts of continuity and determinism, as is evident in his correspondence with Pauli and Einstein. A letter to Pauli of 23 Dec 1919 contains

"One should not transfer the concept of space-time as a four-dimensional continuum from the macroscopic world of common experience to the atomistic world; manifestly the latter requires a different type of manifold". ("Man darf die Begriffe des Raumes und der Zeit als ein 4-dimensionales Kontinuum nicht von der makroskopischen Erfahrungswelt auf die atomistische Welt übertragen, diese verlangt offenbar eine andere Art von Mannigfaltigkeit als adäquates Bild").

Born's correspondence with Einstein from 1920 onwards reflects Born's conviction that the space-time continuum is no longer applicable on atomic scales. But Einstein is not to be convinced; e.g. Einstein's letter to Born of 27 Jan 1920 contains: "I do not believe that the continuum has to be abandoned to solve the quantum problem." ("Daran, dass man die Quanten lösen müsse durch Aufgeben des Kontinuums, glaube ich nicht"). Einstein's insists that the fundamental laws of nature should rely on differential equations. He admits that his own attempts to formulate such equations were unsuccessful; but he will continue to pursue this path.

Born, however, was convinced that a radical breakaway from classical concepts was necessary. It should take several years until this conviction led to a valid theory. The concept of discontinuous and statistical "*quantum transitions*" ("*Quantensprünge*") grew slowly during several years until Born, Heisenberg, and Jordan developed the elementary quantum theory during the year 1925. Einstein's own papers constituted the primary source for Born's conviction. In 1905 Einstein introduced light quanta, which could only be absorbed and emitted as indivisible entities; in 1917 his paper "Zur Quantentheorie der Strahlung" introduced transition probabilities. Einstein himself wanted them to be understood as elements of a thermodynamic description of thermal equilibrium between matter and radiation; he remained convinced that the elementary processes should obey deterministic laws in principle. Born, however, made the decisive step further, he postulated discontinuous and statistical quantum transitions.

The publication "Quantentheorie und Störungsrechnung" (M. Born, Naturwissenschaften 11, 537–542, 1923) confirmed Born's conviction that a radical change was required away from classical concepts prevalent in the old quantum theory. The old quantum rules had led to success only for the simplest possible two body problem coupling a positively charged nucleus with one single electron, but all attempts for a satisfactory treatment of the Helium atom, a three body system, had resulted in failure. Born concludes:

"It becomes more and more probable, that not only new assumptions in the usual sense of physical hypotheses are needed, but that the entire system of basic concepts in physics will have to be rebuilt radically." ("Es wird immer wahrscheinlicher, dass nicht nur neue

Annahmen im gewöhnlichen Sinne physikalischer Hypothesen erforderlich sein werden, sondern dass das ganze System der Begriffe der Physik von Grund auf umgebaut werden muss.")

The general direction which Born had in mind at this time about how to accomplish this radical reconstruction may be inferred from the letter Heisenberg sent to Pauli on 9 Oct 1923. Heisenberg had just arrived in Göttingen to take up the post of Born's Assistent, Pascual Jordan was still completing his doctoral thesis (the doctoral exam taking place in the spring of 1924). Heisenberg writes: "Born specifies our aim for the time ahead: discretization of Atomic Physics" ("Born fasst unsere Aufgaben für die nächste Zeit mit dem Wort zusammen: Diskretisierung der Atomphysik").

Explicit details were contained in the lectures on "Atommechanik", which Born held during the winter semester 1923/24. The lectures explored the limits of the old quantum theory and contained the program how to go beyond these limits towards the "endgültige Atommechanik" (final atom mechanics). The lecture notes were published in the book 'Vorlesungen über Atommechanik, 1. Band' (Lectures on Atom Mechanics, 1st Volume); Springer Verlag, November 1924. The future 2nd volume, should (so the very optimistic announcement) contain the "final atom mechanics".

Born accepts the validity of classical mechanics and electrodynamics for **macroscopic processes** only, and recognizes their failure for the understanding of the quantum world. He criticizes that the old quantum theory conserves concepts of classical mechanics, in particular the continuous movements of electrons, which are not accessible to observations ("Es scheint, dass diese Grössen prinzipiell der Beobachtung nicht zugänglich sind".) Born concludes that the old quantum theory provides a formal calculational scheme, applicable to special cases only, but does not contain the "true quantum laws". He specifies that **the true quantum laws should contain relations between quantities, which are observable** ("Von diesen wahren Gesetzen müssen wir verlangen, dass sie nur Beziehungen zwischen beobachtbaren Grössen enthalten").

The path towards the future Quantum Theory is defined as

"the systematic transformation of classical mechanics into a discontinuous atomic mechanics.....

the new mechanics replaces the continuous manifold of (classical) states by a discrete manifold, which is described by "quantum numbers"....

quantum transitions between different states are determined by probabilities...

the theoretical determination of these probabilities is one of the profound tasks of Quantum Theory....".

This program was conceived before the successful development of the theory; and it is quite logical that the mathematical formulation developed by Born, Heisenberg and Jordan of the new fundamental laws will have discontinuous form, matrix mechanics.

The first concrete step to abandon the space-time continuum and replace it by a new type of manifold is contained in the paper

**Max Born, "Über Quantenmechanik" (Z. Phys. 26, 379–395, 1924)**.

Born's line of attack on the quantum problem is essentially different from the old quantum theory: Whereas the Bohr-Sommerfeld approach used Planck's constant to quantize energies of stationary states, leaving the transition processes still open, Born directly attacked the dynamic behavior, i.e. the **quantum dynamics** describing transitions between different states. It is in this publication that Born introduces the term '**Quantenmechanik**', which indicates that the mechanical behavior is not continuous on the elementary level; all changes proceed in quantized, i.e. discontinuous, steps. Born starts from quantization of action, which he accepts as established by experiment. Born's central argument will be, that quantization of action in terms of Planck's constant $h$ will change the continuous variation of classical variables into **discontinuous quantum transitions**, requiring all physical variables to change discontinuously.

To put this idea into mathematical form, Born starts from classical mechanics using the Hamilton-Jacobi formalism, where action can be used as physical variable, its canonically conjugated variable being a dimensionless angle variable. Detailed knowledge of the Hamilton-Jacobi formalism was not commonplace at the time (nor today), and its use made Born's publications difficult to read. Born had obtained his Doctorate and Habilitation in mathematics and he used the mathematical technique most useful for the physical problem to be solved; action as physical variable was particularly suited, because the central idea, "**changes of action variables are restricted to integer multiples of** $h$", could be implemented directly.

The second mathematical technique, essential for the implementation of Born's concept, is perturbation theory. Classical laws describing dynamics typically are formulated in form of differential equations, where infinitesimally small variations of physical variables are related to each other. Born wanted to replace infinitesimally small classical intervals by their discontinuous quantum equivalents. To achieve his goal, he started from classical lowest order perturbation theory, applicable to infinitesimally small changes; the transition from continuous classical dynamics to discontinuous "quantum dynamics" is achieved requiring action intervals to be multiples of the finite quantum of action $h$. Infinitesimal changes of other physical variables in classical physics are expressed via their dependence on the action variables and differential quotients are replaced by difference quotients. Classical laws in terms of differential equations are thereby replaced by quantum mechanical difference equations.

Let us summarize the central idea of Quantenmechanik by

**Born's quantization condition:**
**All elementary changes occurring in nature must be discontinuous, because the action variables may change by integer multiples of Planck's constant only; the discrete behavior of action variables will affect all other variables as well.**

Although Born's first publication "Über Quantenmechanik" did not carry the discretization of nature very far, a complete elementary Quantum Theory followed during the following year, developed in a series of four publications discussed below.

## 5.2   M. Born and P. Jordan: "Zur Quantentheorie Aperiodischer Vorgänge"

The decisive step, which defines the direction towards a valid Quantum Theory, is contained in the publication by **Max Born and Pascual Jordan (Z. Phys. 33, 479–505, 1925; rec. 11 June 1925)**, referred to as **BJ-1** in the following.

Born recognized that Quantum Mechanics without Quantum Optics is logically inconsistent; quantization of the action variable has to affect all physical variables and all laws of nature. Born's conviction was strengthened by new experimental information, which had become available. Coincidence experiments by **W. Bothe and H. Geiger (Z. Phys. 32, 639, 1925, rec. 25 April 1925)** had shown that Einstein's photon concept requiring momentum and energy conservation for all elementary processes gained strong support from their experimental results on the Compton Effect. These results were known in Göttingen and Copenhagen before submission, and it was recognized that the theory of Bohr, Kramers and Slater (**"BKS"**) was thereby invalidated. Born had been informed about preliminary results already in January 1925; Bohr was informed by letter from Geiger on 21 April 1925, containing the essential results. On the same day Bohr himself sent a letter to James Franck, Born's colleague in Göttingen. Franck showed the letter to Born, which prompted Born to inform Bohr (letter of 24 April 1925) about his own activities concerning the interaction of light with matter. Born writes that, together with Pascual Jordan, he is preparing a publication which considered transitions between stationary states as truly instantaneous quantum jumps, emitting or absorbing quanta of light, thereby adopting Einstein's concept of quantized radiation.

It is in this paper of June 1925 (BJ-1)[1] that Born's quantization concept is extended from Quantum **Mechanics** to include Quantum **Optics**. The introduction of Quantenmechanik by Born had been based on quantization of the action variable, requiring discontinuous dynamics; but this same reasoning has to affect not only mechanical variables but the radiation field as well. Born and Jordan introduce the expression '**Quantenoptik**', attributing its basic laws to Einstein. Born's intention to transform continuous classical mechanics into a discontinuous Quantum Mechanics logically had to be compatible with Einstein's Quantum Optics.

A general principle is formulated, which serves as guide on the path from classical to quantum behavior: **"A fundamental principle of wide range and fruitfulness states that the true laws of nature contain only such quantities which can be observed and determined in principle."** Relativity theory is referred to as an example for the successful application of this principle. Due to the finite velocity of light it is impossible in principle to define simultaneity of events at different locations; this recognition provided the basis for Einstein's derivation of relativity theory. When Born claimed that the concept of a space-time continuum cannot be applied to atomic phenomena, his motivation was similar to Einstein's reasoning. In order to define a precise location and time in a space-time continuum, measuring rods

---

[1] not to be confused with the paper of September 1925, later called BJ-2.

and clocks are needed, which themselves consist of atoms. Absolute precision on atomic or subatomic scales is therefore unobservable in principle. The observability principle already provided the motivation for Born's paper of 1924 introducing Quantenmechanik; unobservable continuous changes of physical variables were replaced by discontinuous and statistical quantum jumps.

Born and Jordan derive **Quantum mechanical perturbation theory** for radiative transition processes, i.e. for spontaneous emission and induced emission and absorption. The starting point is Einstein's paper "Quantum Theory of Radiation" of 1917, which, as is stated explicitly, contains **"the basic law of Quantum Optics"** ("Grundgesetz der Quantenoptik"). Born and Jordan proceed in two distinct steps:

They start by **classical** perturbation theory, calculating the effects on atomic properties induced by a time dependent classical field of general time dependence. The discussion is thereby extended to aperiodic processes, as indicated in the title of the paper. The coupling between electromagnetic field and atoms, responsible for emission and absorption of radiation, is restricted to the atomic dipole moment. The transition from classical theory to quantum mechanical perturbation theory is obtained by imposition of Born's discrete kinematics and dynamics "by hand": The classical mechanical variables are replaced by a discrete manifold of complex **"quantum vectors"** (which we shall later call **"matrix elements"**), discontinuously connecting two quantum states, whose action variables $J$ differ by a finite multiple $\tau$ of Planck's quantum of action. Concerning the radiation field, discontinuous transitions require emission or absorption of radiation quanta.

To make the connection to the following publications easier, specifications concerning the notation used by Born and Jordan are helpful. For the problem of radiative transitions treated in this publication BJ-1 and the dipole approximation used, the relevant atomic degree of freedom is the electric dipole moment **p**. Born and Jordan use a Fourier expansion of **p** and the Fourier coefficients of **p** are denoted by **A**.[2] The "quantum vectors" defined by Born and Jordan are the Fourier coefficients **A**, which, according to Einstein's 'basic law of quantum optics', fulfill the relation

$$\mathbf{A}_\tau^+(J) = \mathbf{A}_\tau^-(J + \tau h). \tag{5.1}$$

$J$ is the action variable, $h$ is Planck's constant and $\tau$ is a positive integer. $\mathbf{A}_\tau^+(J)$ is a quantum vector representing a transition increasing the action variable from $J$ to $J + \tau h$, and $\mathbf{A}_\tau^-(J + \tau h)$ is a quantum vector representing a transition decreasing the action variable from $J + \tau h$ to $J$. The continuous changes of classical action variables $J$ are discretized, requiring changes to occur in finite multiples of $h$ only. Other classical variables $f$ are expressed via their dependence on the action variable $f(J)$ and are set equal to the average $\frac{1}{2}(f(J \pm \tau h) + f(J))$; their changes are required to be given by finite intervals $(f(J \pm \tau h) - f(J))$.

The discretization and averaging process introduces statistical elements into the new quantum theory. Remark that the **difference** in action variable is quantized:

---

[2]Born and Jordan use gothic letters to indicate vector quantities, as was usual in Germany at the time.

$\Delta J = \pm \tau h$; no other physical variable is quantized. Transitions with finite $\Delta J$ require physical variables $f(J)$ to change by discrete amounts $\Delta f = f(J \pm \tau h) - f(J)$, but the $f(J)$ themselves are not equal to some quantized value. In particular, quantized energies, essential elements of the old quantum theory, are no longer required.

The quantum theoretical probabilities for spontaneous and induced transitions are derived. The guiding principles are that the laws of classical optics must be recovered for averages; this leads to the requirement:

**Transition probabilities are proportional to the absolute square of the "quantum vectors"**. Quantum theoretically "time" as continuous variable has disappeared, being contained implicitly only in "transition probability per unit time".

Energy conservation for individual transitions is respected, consistent with Einstein's photon picture, the photon carrying an energy $h\nu$. The results for absorbed and emitted radiation energies calculated classically are recovered for the averages.

## 5.3 Werner Heisenberg: "Über Quantentheoretische Umdeutung Kinematischer und Mechanischer Beziehungen"

**Heisenberg's "reinterpretation** ("Umdeutung") **paper"** (Z. Phys. 33, 879–893, 1925; rec. 29 July 1925) makes the next important step towards the discontinuous quantum theory, following the direction defined by Born-Jordan, but partly retaining Planck-Bohr-Sommerfeld type ideas and methods. Heisenberg was strongly influenced by Bohr's thinking, he had spent the time from September 1924 in Copenhagen with Bohr, returning to Göttingen in April 1925. The development of Heisenberg's ideas during the critical period between 1924 and 1927 may be traced in his frequent correspondence with Pauli[3] (W. Pauli, Scientific Correspondence, Springer Verlag). After his return to Göttingen in April 1925, Heisenberg witnessed the development of the paper by Born and Jordan (BJ-1).

Heisenberg's important new idea: To use the quantum vectors (= amplitudes) of Born and Jordan directly for calculations and to use **matrix multiplication rules** for these amplitudes. Following Born and Jordan, Heisenberg states the intention to obtain quantum laws containing observable quantities only. The second important part taken from Born and Jordan's paper is the partly discontinuous representation of mechanical degrees of freedom. But Heisenberg also retains the essential elements of Bohr's concept. In Bohr's view radiative transitions between two stationary states were supposed to be triggered by something vibrating inside the atom (Bohr's virtual oscillators), emitting and absorbing **continuous** radiation at the oscillation frequencies. Similar to Bohr, Heisenberg still considers radiation to be classical; oscillations continuous in time are required and remain to be the essential part of Heisenberg's

---

[3] Both had been students of Sommerfeld to obtain their doctorates in Munich, and both went to Göttingen afterwards to work with Born.

physical picture and interpretation. The primary objective consists in quantization of the hypothetical "virtual oscillator", believed to produce continuous radiation.

Heisenberg addresses radiative transitions, in particular he wants to calculate the intensities of spectral lines. In his opinion the intensities will be determined by the amplitudes of the virtual oscillation process. The mathematical representation of the oscillator is a mixture taken from Born-Jordan concerning discontinuous spatial degrees and Bohr's concept, retaining continuity in time.

The paper starts with general remarks about the coupling of the radiation field to atomic degrees of freedom via dipole and higher multipoles, which leads to the question, how to represent products of classical variables by appropriate quantum theoretical quantities. Changing slightly the notation of Born and Jordan, Heisenberg adopts their quantum vectors $\mathbf{A}(n, n - \alpha)$ to indicate discontinuous spatial variables. Continuity in time is contained in a multiplying factor, depending harmonically on time: $\mathbf{A}(n, n - \alpha)e^{i\omega(n,n-\alpha)t}$. The $\omega(n, n - \alpha)$ are given by $\frac{1}{\hbar}$ $(W(n) - W(n - \alpha))$. The $W(n)$ will become the energies of stationary states. Heisenberg's complex amplitudes vectors $\mathbf{A}(n, n - \alpha)$ correspond to the quantum vectors of Born and Jordan representing the Fourier amplitudes of the electric dipole moment; $n$ and $(n - \alpha)$ characterize different quantum states, Heisenberg's $\alpha$ corresponds to $\tau$ in the paper of Born and Jordan. Heisenberg's technical novelty consists in the use of matrix multiplication rules for the amplitudes; due to the special form of the frequencies $\omega(n, n - \alpha)$, matrix multiplication rules apply to the amplitudes directly, irrespective of the time dependent factor.

The actual calculations carried out are for a single oscillator degree of freedom. Heisenberg represents the position variable by spatially discrete amplitudes depending on time in continuous fashion: $a(n, n - \alpha)e^{i\omega(n,n-\alpha)t}$. These amplitudes are inserted into the classical oscillator equation $\ddot{x} + \omega_0^2 x + \lambda x^3 = 0$; the nonlinear term will be treated in perturbation theory. Stationary states are selected by a modified Bohr-Sommerfeld quantization recipe, relying on continuity in time: The integral of $m\dot{x}^2$ over a period in time is required to be equal to Planck's constant. The modification is due to W. Thomas (Naturwissenschaften 13, 627, 1925) and W. Kuhn (Z.Phys. 33, 408, 1925) in the attempt to describe dispersion; two separate oscillators are used for transitions upwards and downwards in energy, contributing with opposite sign to the quantization integral. The solution determines energies and amplitudes. Concerning the energies obtained, the Thomas-Kuhn modification of the Bohr-Sommerfeld quantization recipe leads to the correct quantum mechanical eigenvalues of the linear oscillator, including the zero point energy.

Heisenberg's method retains the essential defects of the old quantum theory: The Bohr-Sommerfeld quantization determines stationary states corresponding to classically periodic states only, a general quantization scheme is not available. The determination of stationary states corresponds to static properties, the dynamics of transitions between different stationary states is not addressed. More generally, "quantum theoretical equations of motion", i.e. the equivalent of the classical equations of motion to describe quantum dynamics of transitions, are absent. Radiation is still believed to be classical. Relying on Bohr's concept of hidden oscillators emitting continuous radiation, Heisenberg makes the erroneous assumption that the virtual oscillator dynamics generates radiation at the corresponding frequency. Although Heisenberg does not treat the coupling between atomic degrees of freedom and the electromagnetic field explicitly, he nevertheless argues that the squares of the oscillator amplitudes directly determine the intensities of spectral lines.

Heisenberg himself was well aware of the limitations, as his letter to Pauli of July 9, 1925 (just before his paper was submitted) testifies: He is convinced that the critical part (expressing the opinion that electronic orbits are not observable, whereas proper quantum laws should contain observable quantities only), is fully justified, but that his "positive part" (his explicit treatment of quantum oscillations) is rather formal and scanty, but could serve others to proceed further: ("...dass ich aber den positiven für reichlich formal und dürftig halte; aber vielleicht können Leute, die mehr können, etwas Vernünftiges draus machen").

## 5.4   Max Born and Pascual Jordan: "Zur Quantenmechanik"

During the following months **the final breakthrough** is achieved: **M. Born and P. Jordan (Z. Phys. 34, 858–888, rec. 27 Sept 1925), ("BJ-2") obtain the new fundamental laws: commutation relations and "quantum equations of motion"**.

Let us compare the methods and results of Born and Jordan in the preceding publication BJ-1 to those of Heisenberg. In BJ-1 Born and Jordan **do** address quantum dynamics of discontinuous quantum transitions, and they **do** have a generally valid quantum condition, Born's quantization condition: Changes of action variables are restricted to integer multiples of the quantum of action $h$, implying that all physical variables may change in discontinuous steps only. Nevertheless the implementation of their method (i.e. starting from classical equations and discretizing "by hand") is awkward and would have to be adapted to every new problem treated. The aim to achieve a "final quantum theory" must be the discovery of generally applicable quantization conditions and quantum dynamical equations of motion, directly addressing the problem of quantum transitions. The hint from Heisenberg's paper towards this goal consists in the use of matrix multiplication rules for the quantum vectors (= amplitudes = matrix elements) introduced by Born and Jordan in BJ-1. But Heisenberg's continuous time behavior has to be eliminated. Time as continuous variable is inconsistent with discontinuous spatial behavior and discontinuous quantum transitions in general; discontinous processes may contain time only implicitly in transition rates.

In their paper BJ-2, Born and Jordan restrict the discussion of mechanical variables to a single degree of freedom; the generalization to arbitrary degrees of freedom is announced for a succeeding publication, which will follow shortly ( Born, Heisenberg, Jordan, Z. Phys. 35, 557, 1926; rec. 16 Nov. 1925, referred to as **"BHJ"**). Similarly, in BJ-2 the treatment of the coupling to the electromagnetic field, **including field quantization**, is derived for the coupling to a single electric dipole only.

Field quantization however, is possible for all modes, due to the linear character of Maxwell's equations, allowing decomposition into linearly independent modes.

The key, which will open the path towards the solution of the quantum puzzle, is the general quantization condition. Born's quantization condition—requiring the action variable to change by integer multiples of $h$ only—is applied to the product of the canonically conjugated variables position $q$ and momentum $p$. The classical variables are replaced by their respective Fourier expansions and the classical Fourier coefficients are replaced by discontinuous matrix elements. Using matrix multiplication rules as suggested by Heisenberg, the classical product of $q$ and $p$ is thereby replaced by the matrix product $\tilde{q}\ \tilde{p}$. Requiring action to change by discrete steps of $h$ only, the general quantization condition is obtained:

$$\tilde{p}\ \tilde{q} - \tilde{q}\ \tilde{p} = \frac{h}{2\pi i}\tilde{1}. \tag{5.2}$$

The classical Hamiltonian equations retain their validity, if the classical variables are replaced by their associated matrices. Using the general quantization condition, Born and Jordan show that the classical equations of motion may be transformed into new quantum mechanical equations of motion

$$\dot{\tilde{q}} = \frac{2\pi i}{h}(\tilde{H}\ \tilde{q} - \tilde{q}\ \tilde{H}), \qquad \dot{\tilde{p}} = \frac{2\pi i}{h}(\tilde{H}\ \tilde{p} - \tilde{p}\ \tilde{H}), \tag{5.3}$$

where $\tilde{H}$ is the matrix representing the Hamiltonian.

Born and Jordan show that for a general physical variable $g$ and its matrix representation $\tilde{g}$, the corresponding equation of motion has identical form

$$\dot{\tilde{g}} = \frac{2\pi i}{h}(\tilde{H}\ \tilde{g} - \tilde{g}\ \tilde{H}). \tag{5.4}$$

In their original form the quantum theoretical equations of motion are **not differential equations**: their right hand sides are the commutators of two matrices; the resulting matrices are the quantum mechanical equivalent of the classical variables $\dot{q}$ and $\dot{p}$, which, like any other physical variable, are represented by matrices. For example, the matrix element $\dot{\tilde{p}}_{ba}$ is defined to be given by

$$\dot{\tilde{p}}_{ba} = \frac{2\pi i}{h}\sum_c [\tilde{H}_{bc}\ \tilde{p}_{ca} - \tilde{p}_{bc}\ \tilde{H}_{ca}]. \tag{5.5}$$

This equation expresses the following physical content: The discontinuous transition from some initial state $|a\rangle$ to some final state $|b\rangle$ causes a particle to undergo a **discontinuous momentum change**, which is expressed by the matrix element $\dot{\tilde{p}}_{ba}$. In classical physics $\dot{q}$ and $\dot{p}$ are defined by quotients of infinitesimally small intervals. According to Born's Quantenmechanik however, the basic quantum laws of nature do not allow infinitesimally small intervals; the replacement of the infinitesimally small intervals of classical physics by the discontinuous matrix element is the direct consequence of this fundamental principle. To avoid confusion: The indices $a$, $b$, $c$ may refer to a continuum of values, the **transitions** from $a$ to $b$ or $b$ to $c$ are discontinuous; all transitions are quantized.

As was already contained in BJ-1, the transition probability is proportional to $| \tilde{p}_{ba} |^2$. The particular form of the Hamiltonian matrix will determine which quantum transitions are possible. The diagonal matrix elements are average values, for example: $\tilde{q}_{aa}$ is the average position in state $|a\rangle$.

Time does no longer appear explicitly in the original form of these equations of motion, but is contained implicitly in the requirement:

**Transition probabilities per unit time are proportional to absolute squares of non-diagonal matrix elements**.

This requirement, the commutation relation, and the new quantum equations of motion represent the basic quantum laws.

The final chapter of BJ-2 contains the coupling to the electromagnetic field, **field quantization**, and the calculation of the radiation energy emitted by an oscillating dipole. Although the method used by Born and Jordan is unusual and will later be replaced by others, the essential elements of field quantization are described. Quantization of the action variable must affect electric and magnetic fields as well, which are represented by associated matrices. Maxwell's equations are retained for matrices representing electric and magnetic fields. Born and Jordan conclude, that the method to solve the classical Maxwell equations may be carried over to the quantum problem for propagation in vacuum. The classical solution is obtained by decomposition of Maxwell's equations into infinitely many noninteracting modes of harmonic oscillators. The quantum behavior of the simple harmonic oscillator may thus be applied to these modes and—since they are noninteracting—to all. Furthermore, the interaction of a single mechanical degree of freedom to infinitely many noninteracting radiation modes may be carried out to lowest order perturbation theory, neglecting backcoupling effects. To calculate the radiation energy emitted, Born and Jordan determine the matrix equivalent of the Poynting vector for a radiating electric dipole; calculating the radiation energy emitted, the classical results are recovered for the averages.

## 5.5 M. Born, W. Heisenberg, P. Jordan: "Zur Quantenmechanik II"

The completion of the elementary Quantum Theory was a common effort by **Born, Heisenberg, and Jordan (Z. Phys. 35, 557, 1926; rec. 16 Nov. 1925)**, which will be referred to as **BHJ**.

Concerning the history of this important publication a few remarks are helpful. Heisenberg was absent from Göttingen, when Born and Jordan worked on and completed their paper BJ-2. He had left in the middle of July to travel via Holland to

England, accepting invitations by P. Ehrenfest to Leyden and R. H. Fowler to Cambridge. Afterwards Heisenberg joined Bohr's group in Copenhagen, returning to Göttingen only in the middle of October. In the middle of August Born informed Heisenberg about the progress achieved. During his Copenhagen visit Heisenberg contributed to the paper BHJ by correspondence. His return to Göttingen in the middle of October left little overlap with Born, who left Göttingen on October 29 for a lecturing tour of the US.[4] Born returned to Göttingen in April 1926.

A first draft of the paper BHJ had been prepared by Born and Jordan before Born's departure. The final version contains large sections rewritten by Heisenberg, who was still very much influenced by the Copenhagen concept. Heisenberg's letters to Pauli of 23 Oct and 16 Nov, 1925, testify of considerable differences in attitude and interpretation between the three authors, and these differences partly remained afterwards (later chapters will contain more details).

This paper by Born, Heisenberg, and Jordan (**"BHJ"**) contains "almost everything" about the formal development of elementary quantum theory:

- Treatment of systems with arbitrarily many degrees of freedom;
- Perturbation theory for non-degenerate and a large class of degenerate systems;
- The relation to the eigenvalue theory of Hermitian forms; discrete and continuous spectra;
- Angular momentum algebra is developed;
- General conservation laws (energy, momentum, angular momentum, where applicable) are derived;
- Quantization of the electromagnetic field, selection rules for radiative transitions, intensities of spectral lines;
- Statistical treatment of black body radiation, quantum statistical derivation of Einsteins fluctuation formula.

In the meantime Heisenberg's visit to Cambridge had generated an important consequence. There Heisenberg had described the ideas of his new paper and had been asked to provide a copy when available. Heisenberg sent a copy to Fowler towards the end of August, who passed it on to his young collaborator Paul Dirac. During the following months Dirac rederived the major results of Born and Jordan: Commutation relations and quantum equations of motion. He submitted the paper "The Fundamental Equations of Quantum Mechanics" on 7 November 1925. (**P.A.M. Dirac, Proc. Roy. Soc. A 109, 642–653, 1925**).

---

[4] After an extensive stay at MIT, further lectures were given at other universities (Chicago, Wisconsin, Berkeley, Cal-Tech., Columbia); thereby the new message about quantum theory arrived in the US very quickly after its conception.

## 5.6 Pauli's Solution of the Hydrogen Problem

In parallel to the conception of BHJ, Wolfgang Pauli succeeded in solving the hydrogen problem using the new quantum theory of matrix mechanics. On 3 Nov. 1925 Heisenberg replies to a letter of Pauli (Pauli's letter is not conserved); Heisenberg expresses his joy about learning of Pauli's solution. The news is spreading quickly, even well before Pauli will eventually submit his paper (**W. Pauli, Z. Phys. 36, 336, 1926, rec. 17 Jan 1926**). On 17 Nov 1925 Pauli sends a letter to Bohr with an extensive description of his results, including the Balmer formula and the results for the Stark effect. Bohr mentions Pauli's solution in publications in Nature (116, 845, 1925, appearing in print on 5 Dec. 1925) and in Naturwissenschaften (14, 1–10, 1926, in print on 1 Jan 1926)

Pauli solves the eigenvalue problem of the one body problem in an attractive $1/r$ potential using the algebraic methods of matrix mechanics. He recognizes that the matrices corresponding to the Hamiltonian, $1/2m\ \mathbf{p}^2 - e^2/r$, the square of angular momentum $\mathbf{L}^2 = (\mathbf{r} \times \mathbf{p})^2$, its z-component $L_z$, and the quantum equivalent of the square of the "Laplace-Runge-Lenz vector", defined by

$$\mathbf{A}^2 = (\frac{1}{e^2 m}\ \mathbf{L} \times \mathbf{p} + \mathbf{r}/r)^2 , \tag{5.6}$$

are all mutually commutative and can be diagonalized simultaneously. Using algebraic and combinatorial arguments Pauli obtains the complete solution.

## 5.7 Differences in Understanding Between Born-Jordan and Bohr-Heisenberg

Pauli and Heisenberg had both been students of Sommerfeld to obtain their doctorates in Munich, and both went to Göttingen afterwards to work with Born. Born had recognized the difference between classical and quantum dynamics (action variables may change by integer multiples of h only) already before Pauli became Born's Assistent in Göttingen in 1921 (see his letter to Pauli of Dec. 19, 1919, mentioned earlier). Born suggested to Pauli to use Hamilton-Jacobi mechanics as method for perturbation theory (a common publication resulted: M. Born, W. Pauli, Z.Phys. 10, 137, 1922), the type of method which Born later used in his 1924 publication (introducing the term Quantenmechanik) and in his book "Vorlesungen über Atommechanik, 1. Band" (Nov. 1924), which were the starting points for the new quantum theory. But Pauli was dissatisfied with Born, he left Göttingen after 6 months already to take a new post in Hamburg; afterwards Pauli typically made derogatory remarks about Born, criticizing Born's heavy mathematics! What Pauli (and even Heisenberg later on) failed to see, was that Born started from experimental observations, which he called the "documents of nature". Born generally maintained close contact with his experimental colleagues. Before coming to Göttingen in 1921, he was at the head of the "Physikalisches Institut" of Frankfurt University, which comprised experimental and theoretical physics. When an appointment for the equivalent position in Göttingen was offered, he managed to obtain a common appointment together with James Franck, who was to lead the Institute for Experimental Physics, Born the one for Theoretical Physics. The collaboration between Born and Franck (as long as it was allowed to last) was the central reason to make Göttingen the leading place in Quantum Physics. It was from experimental results that Born deduced his physical arguments, and, making use of his mathematical expertise, Born then converted the physical ideas into a theory expressed by mathematical formulae.[5]

Heisenberg's first collaboration with Born occurred during the winter semester 1922/23, when Sommerfeld was absent from Munich and had sent Heisenberg to work with Born during his absence. Born was impressed by Heisenberg[6] and offered Heisenberg to become his Assistent after the completion of Heisenberg's Doctorate in July 1923. Pauli's skeptical attitude towards Born influenced Heisenberg, who was looking more towards Bohr for physical concepts and ideas, considering Born too mathematical (see his letters to Pauli of 23 Oct and of 16 Nov 1925). Heisenberg and Pauli were both strongly influenced by and had great admiration for Niels Bohr, who was generally considered to be the highest "quantum authority". But whereas Pauli had very strong personal ideas already[7] Heisenberg adhered rather strictly to the Copenhagen concept.

Effectively Heisenberg may already be considered as a member of Bohr's Copenhagen group during the entire period when matrix mechanics was developed. He had been on leave from Göttingen to Copenhagen since the start of September 1924, returning to Göttingen only at the end of April 1925 to fulfill his lecturing duties during the summer semester (May to July). There he witnessed the completion of Born and Jordan's paper BJ-1 (rec. for publication 11 June 1925). The bulk of Heisenberg's paper submitted in July was carried out during a 10 day long visit of Helgoland in mid June. Before the end of the semester Heisenberg left Göttingen (around July 12) to visit Leyden and Cambridge. Afterwards he went on vacation, then returning to Copenhagen. In the meantime Born and Jordan achieved the final breakthrough in their paper BJ-2 (rec. for publication 27 Sept 1925), containing commutation relations, quantum equations of motion, and quantization of the radiation field. When Heisenberg returned to Göttingen in mid October, Born was about to leave Göttingen

---

[5]The criticism of Born's "heavy mathematics" was partly understandable; occasionally it did occur that the physical content was somewhat buried in the mathematical formulation and difficult to extract.

[6]Concerning Born's attitude towards Pauli and Heisenberg, the letter to Einstein of 7 April 1923 contains: "I had Heisenberg here during the winter.....equally as gifted as Pauli... but nicer and more pleasant".

[7]Pauli's publications on the Compton effect (Z. Phys. 18, 272, 1923, and Z. Phys. 22, 261, 1924) relied on Einstein's light quanta, in disagreement with Bohr's continuous radiation. Pauli's letter of 12 Dec 1924 to Bohr (containing the manuscript of the "Pauli principle", Z. Phys 31, 765, 1925) criticized Bohr's continuous orbits; Pauli expressed his support of Born's concept about discontinuous quantum transitions, characterizing stationary states only by quantum numbers.

for his visit of the US on 29 October. During this entire period of the conception of matrix mechanics Heisenberg's scientific contacts with Born were much less than with Bohr, whose physical concept differed drastically from Born's. The central point of the Born-Jordan approach consisted in the elimination of the classical concepts of continuity in space and time, requiring not only quantized mechanical behavior but quantization of the radiation field as well. This was opposite to the Copenhagen perspective of Bohr. And Heisenberg, like Bohr, considered the contributions of Born and Jordan to provide a mathematical technique only, without recognizing their full implications. Heisenberg's physical comprehension was mainly influenced by Bohr, whose thinking was rooted in classical concepts of continuity in space and time. Stationary states were supposed to contain electron dynamics, where the electron position was varying continuously in space and time. Transition processes between stationary states, although called "discontinuous", should still have finite duration, as Bohr stated explicitly in his review "Die Grundpostulate der Quantentheorie" (Z. Phys. 13, 117–165, 1923). Bohr's explicit rejection of Einstein's light quanta required oscillators emitting continuous radiation during a transition process, which itself had to be continuous and therefore of finite duration. Bohr considered the Göttingen breakthrough to provide a mathematical framework for his own ideas; Heisenberg's understanding was similar.

Bohr- and Heisenberg is drawn towards Bohr's ideas as well—retains the conviction that quantum phenomena should be explained in terms of classical concepts. Field quantization was rejected by Bohr and Heisenberg. Even after the common paper BHJ of November 1925, which includes an extensive chapter on field quantization and the derivation of Einstein's fluctuation formulas, Heisenberg did not accept field quantization. He retained Bohr's idea about continuous radiation, as stated explicitly in a presentation to the German Mathematical Society on 19 December 1925 (published in Math. Ann. 95, 683–705, 1926).[8]

Born's understanding was essentially different. In November and December 1925 he described his criticism of the old quantum theory and the motivation for the new concepts in lectures given at the Massachusetts Institute of Technology.[9] Atomic problems require new concepts, radically different from the old quantum theory: "...weak palliatives cannot overcome the staggering difficulties so far encountered" [in the old quantum theory], "the change must reach the very foundations." Born stipulates a new general principle is required, "a philosophical idea", and he refers to Einstein's reasoning, which led him to the theory of relativity. Einstein recognized that simultaneity of two events occurring in different locations is fundamentally impossible to verify by direct observation. The application of the principle 'Theory must be consistent with what is observable' led to relativity theory. Already in their

---

[8]Field quantization will obtain wider recognition only after Dirac's stay in Göttingen from May to September 1926, from where he moved on to Copenhagen, and Dirac's publication on field quantization (Proc. R. Soc. Lond. A 114, 243–265, 1927) carried out in Copenhagen afterwards.

[9]At MIT Born gave 20 lectures on Quantum Theory, followed by 10 lectures on Lattice Theory. These lectures were written up and published quickly by MIT (Max Born: Problems of Atomic Dynamics). Born's lectures at MIT were followed by visits of other Universities (Chicago, Wisconsin, Berkeley, Cal-Tech., Columbia), where further lectures were given.

paper BJ-1 Born and Jordan had referred to the same principle: "The true laws of nature are relations between magnitudes which must be fundamentally observable." Now Born specifies the criticism of the old quantum theory for having "..introduced, as fundamental constituents, magnitudes of very doubtful observability, as, for instance, the position, velocity, and period of the electron". Born concludes that— quite generally—the concept of exact position and exact time has to be abandoned, "...for in order to determine lengths or times, measuring rods and clocks are required. The latter, however, consist themselves of atoms and therefore break down in the realm of atomic dimensions." The new Quantum Mechanics of Born and Jordan eliminates the space time continuum at the small scales of atomic dimensions; continuum concepts acquire the status of averages only.

After the completion of the new Quantum Theory, the essential differences in understanding, between Born and Jordan on one side and Heisenberg and Bohr on the other, remained. Heisenberg and Bohr were still convinced that stationary states and the calculation of their quantized energies constituted the essential objectives of Quantum Theory. That amounted to what Heisenberg considered to be the "integration of the equation of motion", as stated in his letters to Pauli. Bohr's supposedly stationary states, however, correspond to static properties, there is no quantum dynamics and the transition processes are still open. For Born and Jordan **quantum dynamics** of transition processes, in particular the determination of transition probabilities, constituted the essential aim.

Their differences are illustrated by their diverging understandings of radiative transitions. Atomic energies themselves are not observable; observable quantities result from transitions. In radiative transitions, for example, only the emitted radiation is observable, providing information about energy differences. Theoretical descriptions of transitions may start from atomic models as a closed systems, i.e. without interaction to the surroundings. The new Quantum Theory provided a method to diagonalize the Hamiltonian and obtain eigenvalues and eigenstates. For Heisenberg and Bohr the problem was now finished; some hidden oscillators inside the atom—not described by the model itself—were supposed to generate continuous radiation. For Born and Jordan the diagonalization of the atomic Hamiltonian was only a first preliminary step; the second preliminary step consisted in diagonalization of the Hamiltonian of the free electromagnetic field (free, i.e. without coupling to charges). Transitions, however, had to be caused by the interaction between atom and radiation field. Whereas the possibility to obtain the respective eigenvalues was an important first step, the determination of transition probabilities—obtained in perturbation theory—was the primary objective.

The principle cause for the differences in understanding may be traced to the quantization conditions. The Bohr-Sommerfeld condition quantized energies; Born's condition quantized transitions, changes in action variables occur in multiples of $h$. Born's quantization implies that quantized energies of physical systems do not exist. Mathematical models may describe closed systems, but all physical systems are necessarily open systems; there will always be couplings to "the rest of the world" generating transitions. Quantized dynamics is necessarily linked to statistical consequences. Physical variables within a quantum state can no longer be defined exactly, implying quantum uncertainties. All physical variables are affected; atomic energies, for example, may no longer have perfectly sharp quantized values; the interaction between atoms and radiation field will necessarily lead to quantum uncertainties of atomic energies, reflected in finite linewidths.

The differences in understanding between Born-Jordan and Bohr-Heisenberg will become apparent most drastically in the interpretation of uncertainty relations, which will be discussed extensively in Chap. 7.

## 5.8 Brief Summary of the New Quantum Theory

Let us summarize the new message contained in matrix mechanics: Already several years before 1925 Born had claimed that "the entire system of basic concepts in physics will have to be rebuilt radically". And that is indeed what the new Quantum Theory achieved. All physical variables lost their classical significance, being replaced by an associated matrix; particle position or any other variable no longer have precise values, only statistical statements are possible. Time as well lost its traditional meaning; just as position can no longer be exactly assigned to a particle, time as continuous variable has no meaning for a quantum system. Changes occur discontinuously and statistically; 'time' may only be defined implicitly by 'transition probability per unit time'. The general quantum conditions take the form of commutation relations; the quantum theoretical equations of motion have become relations between matrices; for example the time derivative of momentum in classical physics is replaced by a matrix, which, via the new quantum mechanical equation of motion, is related to the commutator of the Hamiltonian matrix with the matrix representing the momentum.

Physical content and mathematical form of the new Quantum Theory were so dramatically different from all traditional concepts in classical physics, that it is no wonder, that the new basic laws shocked the established scientific community, and considerable resistance was widespread. Determinism and continuity, mathematically expressed via differential equations, had been the foundation of classical physics, and it is maybe not surprising that a return to seemingly familiar concepts appeared soon afterwards.

# Chapter 6
# Continuous Representations of the New Quantum Laws

**Abstract** This chapter contains continuous representations of the new quantum laws. Kornel Lanczos pointed out this possibility, used by Max Born and Norbert Wiener to represent "time" by a continuous variable again. Schrödinger's wave mechanics—representing position as continuous variable—provided a mathematic method more familiar than the unusual algebra. Schrödinger conceived it as rejection of the radically new concepts; he refused to give up the space-time continuum: "*Das räumlich-zeitliche Denken*" (i.e. the mode of thought relying on continuity in space and time) should remain the only acceptable way to conceive of processes in nature and remain to be the basis for the understanding of the laws of nature.

**Keywords** Field theoretical representations · "Time representation" · Hermitian operators · Hamilton Jacobi formalism · Commutation relation for time and energy · Phase waves · "Position representation" · Wave functions · Fermi's golden rule · Probabilistic significance of wave functions · Canonical transformations · Scattering process

## 6.1   Kornel Lanczos: Field Theoretical Representations

**Kornel Lanczos (Z.Phys. 35, 812, 1926, rec. 22 Dec. 1925)** was the first to point out, that the discontinuous form of the new fundamental laws of Quantum Mechanics could mathematically be represented by field theoretical methods, relying on functions of continuous variables. The algebraic eigenvalue equations of Born, Heisenberg, Jordan (for example of the Hamiltonian matrix) can be represented by integral equations such that the eigenvalues of the integral operator yield the inverse of the eigenvalues of the algebraic equation. Similarly differential eigenvalue equations might be used, yielding the same eigenvalues as the algebraic equations. If successful, the integral or differential equations should contain the quantization conditions implicitly. Lanczos stressed, however, that field theoretical representations do not imply physical continuity and Lanczos' publication contains a warning: If all physical processes are indeed discontinuous and all physically relevant quantities are contained in discrete matrix elements, then the field theoretical representations nec-

© The Author(s) 2017
H. Capellmann, *The Development of Elementary Quantum Theory*,
SpringerBriefs in History of Science and Technology,
DOI 10.1007/978-3-319-61884-5_6

essarily contain an additional arbitrariness. Infinitely many different representations using different variables are possible yielding the same eigenvalues; the functions of continuous variables then are nothing more than mathematical auxiliary functions, which may be used to calculate the physical significance contained in averages and transition probabilities.

## 6.2  Linear Hermitian Operators and "Time Representation"

The first partially continuous representation was proposed by **Max Born and Norbert Wiener (Z.Phys. 36, (1926), 174–187, rec. 05 Jan 1926)**, when they introduce linear Hermitian operators acting in an infinite dimensional vector space (later called "Hilbert space"), replacing matrices. The previous matrix elements may be written as

$$\tilde{p}_{ba} = \langle b|\hat{p}|a \rangle, \tag{6.1}$$

where $\hat{p}$ is an operator. The basic laws are formulated for Hermitian operators instead of matrices; for any pair of canonically conjugated variables $q$ and $p$ the general quantization condition is contained in the commutation relation for corresponding Hermitian operators

$$\hat{p}\hat{q} - \hat{q}\hat{p} = \frac{h}{2\pi i} = \frac{\hbar}{i}. \tag{6.2}$$

Similarly the "equations of motion" are formulated for operators

$$\dot{\hat{q}} = \frac{i}{\hbar}(\hat{H}\hat{q} - \hat{q}\hat{H}), \qquad \dot{\hat{p}} = \frac{i}{\hbar}(\hat{H}\hat{p} - \hat{p}\hat{H}). \tag{6.3}$$

Referring to the Hamilton-Jacobi formalism of classical mechanics, Born and Wiener remark that 'time' and 'energy' may be considered canonically conjugated variables, just as any other pair of generalized coordinates and momenta. Time may take the role of generalized coordinate and the negative energy will be the canonically conjugated momentum. Remember that the origin of Born's quantization condition resulted from the same principle applied to the canonically conjugated variables action and the associated dimensionless angle variable. Born and Wiener conclude that the commutation relation for time and energy will have to be fulfilled.[1]

$$[\hat{E}\hat{t} - \hat{t}\hat{E}] = -\frac{\hbar}{i} \tag{6.4}$$

---

[1]As will be discussed later in the chapter on "Quantum Uncertainties", commutation relations require the uncertainties of canonically conjugated variables to be connected; for time and energy this relation couples lifetime and energy uncertainties.

In the previous publications BJ-1, BJ-2, and BHJ time had disappeared as explicit variable, being contained implicitly only in'transition probability per unit time'. Now Born and Wiener introduce a special representation: The time operator $\hat{t}$ is represented by the continuous variable $t$ ("time representation"). To fulfill the commutation relation for time and energy the **energy becomes the operator:** $\hat{E} = -\hbar/i \cdot d/dt$. A differential equation with respect to the variable $t$ is obtained

$$H|u(t)\rangle = -\frac{\hbar}{i}\frac{d}{dt}|u(t)\rangle. \qquad (6.5)$$

The dependence of $|u(t)\rangle$ on the variable $t$ does not describe the continuous variation of the physical state of the system with time; $|u(t)\rangle$ is a mathematical auxiliary object, which may be used to calculate matrix elements and the relevant averages and transition probabilities.

The equations of motion for the operators can now be taken to be differential equations with respect to the continuous variable $t$; formal integration yields

$$\hat{q}(t) = e^{\frac{i}{\hbar}\hat{H}t}\,\hat{q}\,e^{-\frac{i}{\hbar}\hat{H}t}, \qquad \hat{p}(t) = e^{\frac{i}{\hbar}\hat{H}t}\,\hat{p}\,e^{-\frac{i}{\hbar}\hat{H}t}, \qquad (6.6)$$

and $|u(t)\rangle$ takes the form

$$|u(t)\rangle = e^{-\frac{i}{\hbar}\hat{H}t}|u\rangle. \qquad (6.7)$$

## 6.3 Wave Mechanics

Between 27 January and 21 June 1926 Erwin Schrödinger published a series of 5 papers in quick succession, which contained the alternative representation of "wave mechanics". Four of the papers had the title "Quantization as Eigenvalue Problem" (1st to 4th communication); intermediate between communications 2 and 3, the paper "On the Relation of the Heisenberg-Born-Jordan Quantum Mechanics to Mine" established the relationship between matrix and wave representations.. Schrödinger motivates his approach referring to the thesis of **Louis de Broglie**, presented in early 1924 (**"Recherches sur la Théorie des Quanta", Ann. de Physique 3, 22–128, Jan. 1925**), which Einstein had already mentioned, when he applied Bose statistics to discuss ideal gases (Sitz. Berlin Ak. d. Wiss. 10-07-1924, and 08-01-1925). Before describing Schrödinger's publications in more detail, a brief discussion of de Broglie's ideas will be given.

## 6.4 De Broglie: Particles and Associated Phase-Waves

De Broglie's starting point was Einstein's theory of special relativity (Ann. Phys. 17, 891–921, 1905; and Ann. Phys. 18, 639–641, 1905), the equivalence of energy and mass and Einstein's photon concept of quantized radiation. De Broglie suggested that photons are particles with extremely small mass, which he called "light atoms" (*atomes de lumière*). He associated an oscillation process of frequency $\nu$ to these "light atoms" via the relation $mc^2 = h\nu$, with $m = m_0(1 - v^2/c^2)^{-1/2}$, and $m_0$ being the rest mass. The velocity $v$ of the light atom should be extremely close to the maximal velocity $c$, such that in this extreme relativistic regime all possible frequencies may be reached while the velocity changed by immeasurably small amounts only. The rest mass $m_0$ of the light atoms should be so small, that in the entire experimentally accessible region the mass $m$ should still remain to be immeasurably small.

De Broglie associated an equivalent oscillation process, which he called "phase waves", to particles with finite mass, in particular to electrons. These "phase waves" should extend over all of space, energy and momentum of "light atoms" and of electrons, however, should still be concentrated in extremely small regions of space. In this sense, de Broglie retained the particle character of "light atoms" and electrons.

For electronic velocities $v$ (for example in x-direction and measured in some "reference system") the associated oscillation process $sin[2\pi\nu(t - vx/c^2)]$ corresponded to a phase velocity $c^2/v$, which, however, should not represent a physical process; the physical velocity of the particle carrying its energy and momentum should be given by the group velocity, which, in all cases, would be smaller that $c$.

Finally, the oscillation process connected wavelength $\lambda$, velocity $v$, and momentum $p$ via Planck's constant: $h/\lambda = mv = p$.

Based on these hypotheses de Broglie had given a new interpretation for the quantization conditions of the old quantum theory: The length of the periodic orbit of stationary states should be an integer multiple of the wavelength $\lambda$.

## 6.5 Schrödinger's "Position Representation"

Already before the series of five publications introducing wave-mechanics Schrödinger announced his intention to take de Broglie's wave concept even more seriously than de Broglie himself. On 15 Dec. 1925 his paper "Zur Einsteinschen Gastheorie" (**Phys. Z. 27, 95–101, 1926**) was received for publication, in which Schrödinger took issue with Einstein's application of Bose statistics to ideal gases. Schrödinger rejected the new statistics for particles; as far as Einstein's particle concept of quantized radiation is concerned, Schrödinger insisted that an oscillating wave picture should be maintained. The possible excitations of these oscillations should be required to occur in integer multiples of $h\nu$ only, which effectively amounted to applying Bose statistics to wave excitations. Bose's quantized phase space volume in terms of $h^3$ was replaced by an equivalent density of allowed waves per frequency interval (imposed by appropriate boundary conditions). Schrödinger announced to take seriously the de Broglie concept for particles with finite mass as well. Moving particles should be viewed as "a kind of wave crest on top of a world background of wave radiation". Schrödinger effectively left the Bose-Einstein mathematics invariant, just the interpretation was different. Whereas Einstein in 1905 had claimed that electromagnetic radiation—just as matter—consisted of elementary objects having particle character, Schrödinger took the opposite direction; particles with finite mass and radiation should "really" be taken to be waves.

Although Schrödinger will refer to de Broglie and waves throughout the series of 5 publication, his mathematical treatment does not contain any relation to de Broglie's. Schrödinger's first publication of the series (**Ann. Phys. 79, 361–376; rec. 27-01-1926**) does not describe the attempt to derive a wave equation, instead an eigenvalue equation is proposed treating the hydrogen problem, which Pauli had solved shortly before using the new matrix-mechanics. Schrödinger claims to derive the equation

$$\Delta \psi + \frac{2m}{K^2}(E + e^2/r)\psi = 0 \tag{6.8}$$

from the Hamilton-Jacobi partial differential equations. K is to be identified as $\hbar = \frac{h}{2\pi}$ in order to reproduce the Balmer series of hydrogen. Schrödinger's solution of the eigenvalue equation is reproduced in many textbooks on Quantum Mechanics.

The procedure from the Hamilton-Jacobi equations, $H(q, \partial S/\partial q) = E$, to the eigenvalue equation appears to be arbitrary. Schrödinger first replaces the action $S$ by $K ln\psi$, then, instead of solving the resulting differential equations, he introduces a variational procedure

$$\delta \int d^3\mathbf{r} \, [(\partial\psi/\partial x))^2 + (\partial\psi/\partial y))^2 + (\partial\psi/\partial z))^2 - \frac{2m}{K^2}(E + e^2/r)(\psi)^2] = 0 \tag{6.9}$$

to arrive at the eigenvalue equation above. If the procedure seems more plausible as an attempt to work backwards, starting from the eigenvalue equation towards the Hamilton Jacobi differential equations, the question remains open, how Schrödinger obtained the equation $\Delta\psi + \frac{2m}{K^2}(E + e^2/r)\psi = 0$. Schrödinger himself does not give any indications.

Effectively Schrödinger has introduced the "position representation", where the position operator $\hat{\mathbf{r}}$ is now taken to be continuous variable $\mathbf{r}$. The quantization condition, $[\hat{\mathbf{p}}\mathbf{r} - \mathbf{r}\hat{\mathbf{p}}] = \hbar/i$, is fulfilled, the operator $\hat{\mathbf{p}}$ is replaced by the gradient with respect to position: $\hat{\mathbf{p}} = \frac{\hbar}{i}\nabla_\mathbf{r}$. Six weeks later, in his third paper, Schrödinger (**Ann. Phys. 79, 734–756 1926, rec. 18 March 1926**) will make this connection explicitly. Lanczos had pointed out that a differential equation could be formulated containing the quantization condition implicitly, and the transition from the commutation relation $[\hat{\mathbf{p}}\hat{\mathbf{r}} - \hat{\mathbf{r}}\hat{\mathbf{p}}] = \hbar/i$ to the continuous $\mathbf{r}$ representation and $\hat{\mathbf{p}} = \frac{\hbar}{i}\nabla_\mathbf{r}$ appears logical.

In his second publication of the series (**Ann. Phys. 79, 489–527, 1926, rec. 23 Feb 1926**) Schrödinger, besides making a first attempt to propose a wave equation, discards the derivation of his first publication, in particular the substitution $S$ by $K ln\psi$, and proposes a different method, relying on the analogy contained in the Hamiltonian variational principle between optics, based on Fermat's principle, and the principle of least action of mechanics. The analogy involves geometric optics not wave optics and serves to replace the "derivation" of the previous publication to arrive at the eigenvalue equation[2] $\Delta\psi + \frac{2m}{\hbar^2}(E - V)\psi = 0$.

The question of obtaining a proper wave equation still had to be addressed. Schrödinger assumes harmonic behavior in time ($\psi \sim e^{\frac{i}{\hbar}Et}$) with frequency $E/\hbar$ and aiming for a wave equation of second derivative with respect to time, he replaces: $(E - V)\psi = -\hbar^2 \frac{E-V}{E^2}\frac{\partial^2}{\partial t^2}\psi$, to arrive at the "wave equation"

$$\Delta\psi - \frac{2m(E - V)}{E^2}\frac{\partial^2}{\partial t^2}\psi = 0. \tag{6.10}$$

But this equation cannot serve as a proper wave equation, since it contains the energy explicitly, restricting the possible solutions to harmonic behavior. The "wave equation" is equivalent to the time independent eigenvalue equation.

Of particular interest is Schrödinger's first reference to the quantum theory of Born, Heisenberg, and Jordan and the description of Schrödinger's attitude towards the new mode of thought it contains. Schrödinger criticizes the elimination of continuity in space and time; he argues that - from the philosophical point of view - this constitutes total capitulation, which he refuses:"We are not really able to change the modes of thought and if we cannot understand within these modes of thought, then we cannot understand at all." ("Denn wir können die Denkformen nicht wirklich ändern und was wir innerhalb derselben nicht verstehen können, das können wir überhaupt nicht verstehen.") A discussion contrasting the attitudes of Einstein, Bohr, Born, and Schrödinger will be given in a later chapter.

Schrödinger's third publication of the series (**Ann. Phys. 79, 734–756 1926, rec. 18 March 1926**) contains the explicit connection to matrix mechanics and the transition from the commutation relation $[\hat{p}\hat{r} - \hat{r}\hat{p}] = \hbar/i$ to the continuous "**r** representation" and $\hat{p} = \frac{\hbar}{i}\nabla_r$. But Schrödinger discards a full equivalence of the two approaches, insisting on the physical significance of wave functions, which he sees as the essential element; wave mechanics should be understood as an extension and part of classical field theories. Concerning matrix mechanics Schrödinger reaffirms his disagreement about physical content and mathematical formulation; he states that he felt deterred, if not to say repelled, by the apparently "very difficult methods of transcendental algebra and the lack of illustrative clarity"("ich fühlte mich durch die mir sehr schwierig scheinenden Methoden der transzendenten Algebra und durch den Mangel an Anschaulichkeit abgeschreckt, um nicht zu sagen abgestossen").

Schrödinger's forth publication of the series (**Ann. Phys. 79, 734–756 1926, rec. 10 May 1926**) contains the wave mechanical version of time independent perturbation theory (equivalent to matrix mechanical perturbation theory of "BHJ") and its application to the Stark effect of the Balmer series, reproducing the results Pauli had obtained in his matrix solution of the hydrogen problem.

In his 5'th publication of the series (**Ann. Phys. 81, 109–139, 1926; rec. 21 June 1926**) Schrödinger recognizes that the time dependent "wave equation" of the second

---

[2]Schrödinger's use of dimensions is somewhat confusing, he leaves out the mass $m$ in this equation. In the following $m$ will be included.

publication cannot be used as general wave equation, since energy $E$ is contained explicitly.

Schrödinger makes several further guesses at wave equations, which originate from the application of the operator $(\Delta - \frac{2m}{\hbar^2} V)$ to the eigenvalue equation for a second time, leading to an equation containing forth order derivatives with respect to position. Again simple harmonic behavior with frequency $\nu = E/\hbar$ is provisionally assumed (equivalent to: $(\frac{E}{\hbar})^2 \psi = -\frac{\partial^2}{\partial t^2} \psi$) leading to

$$(\Delta - \frac{2m}{\hbar^2} V)^2 \psi = -\frac{4m^2}{\hbar^2} \frac{\partial^2}{\partial t^2} \psi, \tag{6.11}$$

which is then proposed to be the "general wave equation for the field scalar $\psi$".

Two more versions follow, which are obtained from factorizing the previous time independent equation: $(\Delta - \frac{2m}{\hbar^2} V \pm \frac{2m}{\hbar^2} E)(\Delta - \frac{2m}{\hbar^2} V \mp \frac{2m}{\hbar^2} E)\psi = 0$. At this point again provisionally assuming behavior in time $\psi \sim e^{i/\hbar \cdot Et}$, the equations

$$(\Delta - \frac{2m}{\hbar^2} V)\psi(t) = \pm \frac{2mi}{\hbar} \frac{\partial}{\partial t} \psi(t) \tag{6.12}$$

are obtained. Schrödinger proposes that either of these equations might serve as generally valid wave equation.

We recognize that the time representation introduced by Born and Wiener combined with Schrödinger's position representation yields the equation with the minus sign, now called the time dependent Schrödinger equation.

Let us recall that the fundamental laws of Quantum Theory obtained by Born and Jordan contain "time" only implicitly via "transition probability per unit time". In the "time-representation" introduced by Born and Wiener ("time" continuous variable $t$ and energy $E = -\hbar/i \cdot d/dt$) the dependence on $t$ of $|u(t)\rangle$ does not represent the continuous change of the physical state of the quantum system, $|u(t)\rangle$ may only serve to calculate the required "transition probability per unit time" to recover the physical meaning of "time" for the quantum system. Born and Wiener might just as well have used $x$ or any other mathematical variable instead of $t$; the computation of transition probabilities does not depend on which mathematical variables are used. The same arguments apply to Schrödingers wave-functions $\psi(t)$. The determination of the transition rate by standard time dependent perturbation theory leads to "Fermi's golden rule", the mathematical variable $t$ has been eliminated.

## 6.6  Max Born: The Probabilistic Significance of Wave Functions

Very quickly after Schrödinger submitted his version of wave mechanics, **Max Born (Z. Phys. 37, (1926), 863–867, rec. 25 June 1926)** established the relation between wave functions, matrix elements and probabilities. Born and Wiener had been the first to implement a mathematical representation of the quantum laws based on continuous variables, the time representation. The time-energy commutation relation, $[\hat{E}\hat{t} - \hat{t}\hat{E}] = -\hbar/i$, could be satisfied by representing the time operator by a simple variable $t$, which required the energy operator to be represented by the derivative with respect to $t$. Although Born and Wiener introduced commutation relation, $\hat{p}\hat{q} - \hat{q}\hat{p} = \hbar/i$, valid for any two canonically conjugated variables $p$ and $q$, they did not choose a representation for position and momentum operators similar to the relation between time and energy. Instead they chose a—more complicated—integral representation. After learning of Schrödinger's papers, Born quickly realized his oversight; mathematically there are infinitely many canonical transformations, and therefore infinitely many representations possible using differential operators. Born did not hesitate to employ Schrödinger's advantageous representation for the treatment of scattering problems.

Explicitly, Born discussed the scattering of an electron by an atom and he used the position representation to calculate the relevant matrix elements. He describes the process of an incoming electron with momentum $\mathbf{p}_i$ and kinetic energy $\epsilon_i = p_i^2/2m$ being scattered by an atom having initial energy $W_i$. The observable quantities are the incoming and outgoing momenta $\mathbf{p}_i$ and $\mathbf{p}_f$. In the spirit of first order perturbation theory the scattering process is taken to result from a single elementary quantum transition induced by the interaction $V_{int}$ between electron and atom to a final electronic state with momentum $\mathbf{p}_f$ and kinetic energy $\epsilon_f = p_f^2/2m$, the atom undergoing a transition to a final state of energy $W_f$. Born uses Schrödinger's position representation to describe the scattering process; the electron is represented by plane wave functions, the incoming state as $\psi_i = sin(2\pi/\lambda_i\ z)$, the outgoing state as $\psi_f = sin\ (2\pi/\lambda_f\ (\alpha x + \beta y + \gamma z))$. Wavelengths and electron momenta are related by $p = h/\lambda$. Total energy conservation requires the outgoing wavelength to be given by $(h/\lambda_f)^2/2m = W_i - W_f + \epsilon_i$. The transition probabilities are determined to be proportional to $|\langle \Psi_f, \psi_f|V_{int}|\Psi_i, \psi_i\rangle|^2$, where $\Psi_i$ and $\Psi_f$ are initial and final atomic states.

Although Born accepted wave functions as mathematical tool to calculate matrix elements, he insisted that Schrödinger's physical interpretation was incorrect; Born writes:

"We do not get an answer to the question, 'what is the state after the collision?', but only to the question, 'how probable is a given result of the collision?'... based on the principles of our Quantum Mechanics there exists no quantity, which determines the result of the collision for the individual elementary process".

Let us summarize the message from Born's "user manual" for Schrödinger wave functions: Although Schrödinger's position representation may be used to correctly calculate physically relevant matrix elements, the solutions to the Schrödinger differential equations do not have direct physical significance:

**Wave-functions $\psi(\mathbf{r}, t)$ are nothing more than mathematical auxiliary functions to calculate averages and probabilities**.

Schrödinger vigorously opposed this interpretation, he had a very different physical picture in mind; particles should "really" be thought of as waves and classical continuum concepts should also be applicable to the quantum world. E.g. Schrödinger's letter to Wilhelm Wien of 21 October 1926 contains: "...the whole game with waves might as well get lost, if its only use is providing a comfortable calculational help to determine matrix elements."

But Born insists; Abraham Pais (Science 218, 1193–1198, 1982) quotes Born during the fall of 1926 in Göttinger Nachrichten:"It is necessary to drop the physical picture of Schrödinger, which aims at revitalization of the classical continuum theory, to retain only the formalism and to fill that with physical content". Pais also cites Born's letter to Einstein of November 1926: "Schrödinger's achievements reduces itself to something purely mathematical, but his physics is quite wretched (kümmerlich)". Einstein's reply of 4 Dec 1926 was a severe disappointment to Born. Einstein's first reaction (letter to Hedi Born of 3 March 1926) to the new Quantum Theory had been quite enthusiastic: "The Heisenberg-Born ideas are keeping everybody in suspense, the thoughts and reflections of all theoretically interested people. Dull resignation has been replaced by a unique tension." But now Einstein voices his skepticism: "Quantum Mechanics is very imposing. But an inner voice tells me that it is not the real thing ("der wahre Jakob"). The theory delivers a lot, but it hardly gets us closer to the secret of "The Old One". At any rate I am convinced that "He" does not play dice."

## 6.7   Brief Summary

Schrödinger's position representation and the resulting wave functions reduced the unfamiliar form of algebraic equations derived by Born, Heisenberg, and Jordan to the familiar territory of differential equations. This was widely considered to be a major advantage and Schrödinger's wave equations became the method of choice for elementary applications of the new quantum theory. But the technical advantage of Schrödinger's wave mechanics became a major obstacle to the understanding of Quantum Mechanics. Whereas the discrete form of the matrix representation directly indicated the fundamental property of discontinuous quantum physics, Schrödinger's differential equations and the resulting wave functions suggested continuity and determinism. Schrödinger himself interpreted the wave functions to have direct physical significance, attributing "true wave character" to single particles. Schrödinger's aim had been to go back to the classical concepts of continuity in space and time, eliminating discontinuous and statistical quantum transitions and returning to continuity and determinism. This interpretation was adopted by many contemporaries.

Schrödinger's differential equations made quantum theory more accessible for calculational purposes, but the understanding of the principles of quantum physics suffered. This is particularly true for scattering experiments and the attribution of wave properties to particles, which directly touches the problem Einstein had pointed to in 1905, when he claimed that electromagnetic radiation consisted of elementary objects having particle properties.

# Chapter 7
# The Consequences of the Basic Quantum Laws on Wave Phenomena and Quantum Uncertainties

**Abstract** This Chapter discusses the consequences of the basic quantum laws on "wave-like" phenomena and quantum uncertainties. The uncertainty relations are generally attributed to Heisenberg. Heisenberg's justification, however, based on necessary disturbances introduced by the measuring process, will turn out to be erroneous. Bohr's "Complementarity principle", postulating "particle-wave duality", completes the so-called Copenhagen interpretation of Quantum Mechanics. Particle-wave duality will be shown to be a mathematical artifact without physical significance.

**Keywords** Light scattering and X-ray scattering · X-ray diffraction · Duane's quantum condition · Bragg peaks · "Energy representation" · "Momentum representation" · Discrete translational symmetry · Uncertainty relations · Disturbance free measurements · Complementarity principle · Dual properties

## 7.1 The Solution to Einstein's Problem: How to Connect Particle Properties and Wave Phenomena

Let us recall how the spectral decomposition of light to determine the Planck spectrum had been performed experimentally. Light was scattered off artificially grated surfaces in grazing incidence. The observation of special reflection maxima and minima was interpreted as a measurement of the wavelength of the incident waves, due to constructive and destructive interference when scattered from the periodic grating. Using the relation $\nu = c/\lambda$ the frequency was determined. Radiation reflected from thin plates was interpreted in similar fashion. In 1912 Max von Laue (Sitzungsberichte der Math.-Phys. Akademie der Wissenschaften (München) 1912; 303) discovered X-ray diffraction by crystals, which was generally considered to prove that X-rays were just another form of electromagnetic radiation of smaller wavelengths. The occurrence of diffraction peaks was interpreted to result from constructive interference of waves reflected from parallel lattice planes. Wave characteristics of light and X-rays—such as wavelengths and frequencies—are not observable directly; the

© The Author(s) 2017
H. Capellmann, *The Development of Elementary Quantum Theory*,
SpringerBriefs in History of Science and Technology,
DOI 10.1007/978-3-319-61884-5_7

assumed wave properties result from indirect interpretations invented to "explain" the observed scattering maxima and minima.

Already before a valid quantum theory was available, **William Duane (Proc. Nat. Ac. Sci. 9, 158–164, 1923)** suggested that phenomena observed in light scattering and X-ray scattering, which had been interpreted to result from constructive and destructive interference of waves, could be understood as a pure quantum phenomenon based exclusively on particle characteristics of photons. Duane suggested that these phenomena are due to a quantum condition imposed on the momentum transfers of the scattered photons. Einstein's particle concept of light was taken to be correct; photon characteristics were energies $\epsilon$ and momenta $\epsilon/c$; no wave properties such as wavelengths or frequencies were required. Duane's arguments were based on a simple dimensional analysis: The selection of special momentum transfers $\Delta p$ should result from a quantum condition involving Planck's constant $h$. To arrive at an equation, a characteristic lengthscale had to be introduced, which–for diffraction phenomena—must be the periodicity length $d$ of the grating or the crystal, leading to $d \cdot \Delta p = nh$. Quantum theory (and experiments when the appropriate experimental techniques became available) proved Duane to be right!

In 1926 Born's scattering paper attributed to Schrödinger's wave-functions the role of mathematical tools to calculate transition probabilities only, without further physical significance. At that time single particle scattering could not yet be resolved experimentally; even if scattering of single photons remains to be difficult today (but feasible), such experiments have become standard practice in neutron scattering. The observation of interference patterns—such as Bragg peaks—building up from a large enough number of independent single particle scattering events is possible in many laboratories. To avoid the misconception that wave functions and "dual properties"—particle like and wave like—of photons (or other particles such as electrons or neutrons) are necessary for the understanding of so-called interference phenomena, mathematical representations may be used, which describe the observable quantities by real numbers. Following the development of the new quantum theory by Born, Heisenberg, Jordan, and Wiener, different representations of the fundamental laws were proposed by Pauli, which are particularly useful to describe scattering processes.

On 14 Jan 1926 Heisenberg sent copies of the Born-Wiener paper and of Lanczos paper to Pauli, who quickly understands the connection between commutation relations and continuous representations. Born and Wiener had used the time-energy commutation relation to introduce the time representation, where time $t$ is continuous variable and the energy becomes the operator $-\hbar/i \cdot d/dt$. Pauli's reply to Heisenberg of 31 Jan 1926 mentions the possibility of an equivalent "energy representation", where energy remains to be real variable $E$, and time becomes the operator $\hbar/i \cdot d/dE$.[1] Pauli's letter of 19 Oct 1926 to Heisenberg contains the definition of

---

[1]Pauli will revert his position about this point later. In his book "Die allgemeinen Prinzipien der Wellenmechanik", Springer Verlag 1933, Pauli concluded that the existence of discrete spectra is

the "momentum representation"; momentum is represented by the real variable $\mathbf{p}$, and the commutation relation requires position to be represented by the gradient with respect to momentum: $\hat{\mathbf{r}} = -\hbar/i \, \nabla_{\mathbf{p}}$. The momentum representation is particularly useful to describe the solution to Einstein's problem; in scattering experiments, initial and final momenta are the measured quantities and their representation by real variables makes the connection between experimental observations and physical content clearer.

Detailed discussions of scattering processes and their connection to the basic laws of Quantum Theory will be given in the appendix. Here the discussion is restricted to the basic elements necessary for the understanding of so-called interference phenomena, which result from **purely elastic scattering processes**. We consider the scattering caused by a static potential $V(\mathbf{r})$. The momentum representation will be used to describe a single particle scattering process; the particle momentum is represented by the real variable $\mathbf{p}$, the position is represented by the operator $\hat{\mathbf{r}} = -\hbar/i \, \nabla_{\mathbf{p}}$. The transition for a particle with initial momentum $\mathbf{p}$ to a state of final momentum $\mathbf{p} + \hbar\mathbf{q}$ may mathematically be expressed by

$$e^{-i\mathbf{q}\hat{\mathbf{r}}}|\mathbf{p}\rangle = |\mathbf{p} + \hbar\mathbf{q}\rangle . \tag{7.1}$$

For a scattering potential $V(\hat{\mathbf{r}}) = \int_{\mathbf{q}} \tilde{V}(\mathbf{q}) \, e^{-i\mathbf{q}\hat{\mathbf{r}}}$ the probability for the particle to be scattered with momentum transfer $\Delta\mathbf{p} = \hbar\mathbf{q}$ will be proportional to $|\tilde{V}(\mathbf{q})|^2$. For a structure which is translationally invariant if displaced by a a set of real space vectors $\mathbf{d_i}$, finite Fourier components are restricted to $\mathbf{q} \cdot \mathbf{d_i} = 2\pi n$. The possible momentum transfers have to fulfill the condition $\Delta\mathbf{p} \cdot \mathbf{d_i} = nh$, as suggested by Duane in 1923. The selection of the allowed momentum transfers is a direct consequence of Born's quantization condition; all quantum transitions require the action variable to change by integer multiples of $h$. For the scattering processes discussed here, the product of a vector $\mathbf{d}$, representing a discrete translational symmetry, and the allowed momentum transfers $\Delta\mathbf{p}$ correspond to the change in action variables of the transitions, which—as required—are integer multiples of $h$.

The second condition to be fulfilled is energy conservation. The special momentum transfers depend on the type of particle to be scattered only in so far as its relation between energy and momentum is concerned. This "dispersion relation" will be different for photons, neutrons, or electrons; but the possible momentum transfers are solely imposed by the periodicity of the scattering potential; transferred momenta will have to be identical for all types of particles: photons, electrons, neutrons,......

The results above may also be applied to other geometries defining the scatterer. Fourier decomposition and restriction to lowest order effects—equivalent to the scattering process being due to a single, purely elastic, elementary quantum transition—allows the application to each Fourier component separately. The geometry of the

---

(Footnote 1 continued)
incompatible with the energy-time commutation relation. Pauli erroneously claimed that time $t$ always has to be treated as an ordinary number. The following section on "Quantum Uncertainties" will contain a detailed discussion of this point.

scatterer determines which momentum transfers are possible, independent of the type of particle to be scattered; thereby all types of diffraction phenomena are covered, double or multiple slit experiments are described just as any other geometry.

Thermal radiation consists of statistically emitted particles without any wave characteristics. The spectral decomposition obtained in a grating spectrum constitutes a decomposition according to different photon momenta. Incoming photons scattered elastically obtain the same momentum transfer and different initial momenta are reflected under different angles. Constructive and destructive interference of single photons or other particles is a mathematical artifact of the position representation. It does not correspond to any real physical process; experimentally observable are particles scattered according to the statistical laws of quantum theory.[2]

A clear distinction is to be made between electromagnetic waves described by Maxwell equations and their elementary constituents, the photons. Maxwell theory describes **vector fields**, which themselves have direct physical significance and are measurable. The electromagnetic waves (e.g. radiated by a linear radio transmitter) require emission of a macroscopic number of photons; this by itself is not yet sufficient to define the macroscopic vector fields; their existence requires the polarization of the emitted photons to be correlated. The polarization structure is imposed by the alternating current of the transmitter, i.e. the source; the correlated polarization of a macroscopic number of photons is necessary to induce an alternating current in an antenna with measurable frequency and wavelength. But the elementary constituents, the quanta, do not have the properties of the macroscopic averages; single photons do not have any wave properties, they are simply particles. Already in 1909 Einstein had pointed to the vector character of the electromagnetic fields. He suggested that a macroscopic number of photons should be necessary to define the fields and the phenomenological equations describing the macroscopic averages should coincide with Maxwell's equations.

Let us recall the conditions imposed on the scattering contributions which had been attributed to interference phenomena: The transitions must be purely elastic, not inducing any change in quantum numbers characterizing the scatterer. These purely elastic events occur with finite probability only, scattering events accompanied by transitions within the scatterer contribute with finite probability as well. E.g. scattering contributions from a crystal will contain processes which are accompanied by localized transitions in atoms; these events contribute finite probabilities to a wide range of possible momentum transfers, generating a probability pattern containing a more or less uniform background, in addition to the Bragg peaks resulting from elastic contributions. The appendix will contain further details about the general connection between the basic laws of Quantum Theory and scattering processes.

---

[2]To actually resolve the scattering of individual photons is experimentally difficult (possible today but not at the time). Nowadays reduced intensity and appropriate detectors make the observation of individual photon scattering possible.

## 7.2   Quantum Uncertainties

Already before a valid quantum theory had been developed, Bose's "quantization of phase space" (i.e. particles can no longer be characterized by precise values of position and momentum, only the association with a finite elementary volume of size $h^3$ may be specified) indicated that quantization is connected with quantum uncertainties. When Born and Jordan in June 1925 (BJ-1) replaced continuous variations of classical behavior by discontinuous quantum transitions, quantum uncertainties of physical variables were contained as necessary elements. Quantization of the action variable for all elementary transitions implied an averaging process: Born's quantization was applied to the change in action variable for quantum transitions between states (not to a physical quantity of initial and final states, as in Bohr-Sommerfeld quantization of energies and angular momenta of stationary states). The physical variables of a particular state, however, remained uncertain and could only be associated with averages over corresponding action intervals of size $h$. The uncertainty relations discussed below will show that the basic quantization condition, expressed by the commutation relations, will require all variables of physical systems to necessarily possess quantum uncertainties.

**Heisenberg (Z.Phys. 43, 172, 1927, rec 23 Mar 1927)** drew special attention to uncertainty relations of canonically conjugated variables and their consequences for experimental observations. The development of Heisenberg's ideas leading to this paper may be traced in the frequent exchange of letter's with Pauli. Of particular interest is Pauli's letter of 19 Oct 1926 with reference to Born's publication (Z. Phys. 37, 863, 1926; rec. 25 June 1926) on scattering, transition probabilities and their relation to matrix elements. Born had used Schrödinger's position representation,[3] Pauli pointed out that a momentum representation might be defined as well. Furthermore Pauli remarks that the general commutation relation for canonically conjugated variables ($pq - qp = \hbar/i$) implies that either $p$ or $q$ can be taken to be arbitrarily precise, but only at the expense of increasing uncertainty of the canonically conjugated variable. Pauli considers this to be a "dark point", which will have to be cleared up. Heisenberg reacts (letter of 28 Oct 1926): "I am very enthusiastic, because one can understand the physical significance of Born's formalism much better".

Heisenberg's intention is expressed in the title of his paper: "Über den anschaulichen Inhalt der quantentheoretischen Kinematik und Mechanik" ("On the illustrative content of quantum theoretical kinematics and mechanics"); he wants to give a "illustrative or graphically appealing" understanding ("anschauliches Verständnis") of the commutation relation and the impossibility for arbitrary precision of canonically conjugated variables. This "illustrative understanding", however, uses illustrations and concepts borrowed from classical physics, which are not suited to "explain" quantum behavior. Heisenberg's essential argument for commutation relations and quantum uncertainties relies on the erroneous assumption, that measure-

---

[3] Heisenberg had criticized Born's publication and the use of Schrödinger's wave functions (letter to Pauli of 28 July 1926 [142]).

ments necessarily introduce disturbances (Störungen) to the system to be measured; **quantum uncertainties** of physical variables—**according to Heisenberg**—were to be **caused by measurements**.

Taking the measurement of position, for example, the line of thought goes as follows: If the position of particle A is to be measured, light or X-rays or other particles B have to be scattered off particle A. Heisenberg argues that this scattering process will necessarily disturb, i.e. change, the original state of particle A, in particular it will transfer momentum to the particle A. This disturbance caused by the measuring process of position should—according to Heisenberg—be responsible for a momentum uncertainty of particle A. To obtain a semi-quantitative estimate, Heisenberg invokes an analogy with the optical microscope, where the classical resolution is limited to lengths of order wavelength $\lambda$, and extends this reasoning to the $\gamma$-ray-microscope. At the time when he wrote the uncertainty paper he had become convinced that the coincidence measurements of **W. Bothe and H. Geiger (Z. Phys. 32, 639, 1925)** required momentum and energy conservation in individual scattering processes. Heisenberg concludes that, if $\gamma$-rays are used to measure the position of a particle, the Compton Effect should produce a momentum transfer to the particle of order $\Delta p = h/\lambda$. The product of electron position uncertainty $\Delta q \sim \lambda$ and the momentum uncertainty produced by the measuring $\gamma$-rays should be $\Delta q \cdot \Delta p \sim h$. If $\gamma$-rays are replaced by other particles B as measuring devices, Heisenberg refers to de Broglie and associates a wavelength $\lambda = h/p$ with particles B. The scattering process again should transfer momentum to particle A, inducing an uncertainty yielding the same estimate $\Delta q \cdot \Delta p \sim h$. Higher precision of position requires shorter wavelengths of $\gamma$-rays or higher momenta of particles B, necessarily inducing larger momentum uncertainties of particle A.

Let us first retain what is valid about uncertainties and commutation relations: (a) Quantum uncertainties and commutation relations are connected and are consequences of the same physical principle. (b) Canonically conjugated variables of a physical system cannot both have arbitrary precision; the product of their relative uncertainties is restricted by a lower bound of order Planck's constant.

Heisenberg's "illustrative explanation", however, is fundamentally flawed; his argument,'Observations necessarily cause changes in the system to be measured; these disturbances are responsible for quantum uncertainties', is incorrect: **Measurements without disturbances in the system to be measured are not only possible, but are carried out routinely**. As has been discussed in the previous section **7.1**, Bragg scattering and all similar phenomena, which had been interpreted to result from interference effects, require disturbance free processes; the scattering system remains unchanged.[4] Disturbance free measurements may even be used to **measure** particle positions and their **quantum uncertainties**; further details will be given in the appendix.

---

[4]Mössbauer spectroscopy is another example.

Heisenberg confirms that his own understanding is in contradiction to Born-Jordan; Heisenberg writes: Quantum uncertainties may "according to Born and Jordan be viewed as characteristic and statistical elements of quantum theory in contrast to classical theory". Heisenberg specifies his deviating attitude:

"The difference between classical and quantum theory rather consists in: Classically we may assume the phases of the atom to be determined by preceding experiments. In reality, however, this is impossible, because every experiment to determine the phase will destroy or change the phase of the atom."

And concerning causality Heisenberg adds:

"The sharp formulation of the causality law,'if we know the present exactly, we are able to calculate the future', is not wrong due to the second part of the sentence, but because the precondition is wrong".

This is in sharp contrast to Born's basic principle that all changes in nature are discontinuous and purely statistical. Quantization of the action variable is the key: Discontinuous transitions require the action variable to change by integer values of $h$; physical systems therefore can no longer be characterized by precise values of physical variables. The basic quantum laws, the commutation relations, are consequences of quantization of the action variable (Born and Jordan, Z. Phys. 34, 858, 1925); and the uncertainty relations of canonically connected variables are necessary consequences of the commutation relations. The formal derivation from commutation relations to uncertainty relations is contained in the paper by **E. H. Kennard (Z.Phys. 44, 326, 1927; rec 17 July 1927)** and is given most clearly by **H. P. Robertson (Phys. Rev. 34, 163, 1929)**.

No physical variable of any physical system may be perfectly sharp. Just as the position-momentum commutation relation does not allow a particle to be in an exact momentum eigenstate, the time-energy commutation relation forbids an atom or any other physical system to be in an exact energy eigenstate. Mathematical models of closed systems may have exact eigenvalues and eigenstates; but these models of closed systems, at best, describe approximations. Physical systems are necessarily open systems, nature always provides additional couplings to "the rest of the world". Taking the relation between energy and time as an example: The additional couplings guarantee that every state of a physical system has finite lifetime, generating a natural linewidth and implying that the energy levels cannot be perfectly sharp. The time-energy commutation relation imposes a lower bound on the product of energy and lifetime uncertainties. The "energy-representation" (energy real variable and time becomes derivative with respect to energy, as proposed by Pauli in his letter to Heisenberg of January 1926) is perfectly admissible.[5] Pauli's "correction" in "Die allgemeinen Prinzipien der Wellenmechanik", Springer Verlag 1933, ('time always has to be treated as ordinary number') is in error.

After submission of Heisenberg's paper, intensive discussions with Bohr continued, which induced Heisenberg to write an addendum. Bohr had developed ideas,

---

[5]The appendix contains an example; the energy representation establishes the connection between "time"-dependent perturbation theory and field quantization.

which only partly agreed with Heisenberg's concept, but also contained essential differences. Although Bohr accepted Heisenberg's erroneous argument concerning experiments necessarily provoking disturbances, Bohr advocated that wave properties of particles should be the primary reason for quantum uncertainties. At first Heisenberg remained skeptical; his letter to Pauli of 4 April 1927 contains: "I am quarreling with Bohr whether the relation $\Delta q \Delta p \sim h$ has its origin in the wave or the discontinuity aspect of quantum mechanics. Bohr stresses, that diffraction of waves is essential, I emphasize that the light-quantum theory and the Geiger-Bothe experiment are essential." Schrödinger's waves should only be calculational tools without physical significance. Concerning'particles, not waves' Heisenberg still agreed with Born and Jordan, the important disagreement lies in the origin for discontinuities and uncertainties: Heisenberg saw them as the result of disturbances produced by scattering processes, whereas Born and Jordan maintained that discontinuities and uncertainties are constitutive elements of quantum physics. Heisenberg's letters to Pauli of 16 May and 25 May 1927 indicate that the controversy with Bohr led to severe tensions, which were defused somewhat by the addendum. Heisenberg acknowledges Bohr to have pointed out that the wave aspect, in particular the uncertainty due to finite collimation of the $\gamma$-ray microscope, should play an important role for the uncertainty of position measurements.

## 7.3   Bohr's Complementarity Principle and Dual Properties

Bohr presented his own arguments at the conference held in Como in September 1927, in commemoration of the 100th anniversary of the death of Alessandro Volta. Bohr's lecture "**The Quantum Postulate and the Recent Development of Atomic Theory**" (published in Naturwissenschaften 15, 245–257, 1928) introduced the "**Complementarity principle**", which became the basis of the Copenhagen interpretation of Quantum Mechanics. According to the Complementarity principle, particles should have "dual properties", i.e. particle **and** wave properties. Experimental results observing particles should be due to the "particle property", experiments observing "interference" should be associated with the "wave property".

During the previous years Bohr had always rigorously opposed Einstein's concept of radiation quanta having particle character. In Bohr's view interference of waves was the only possibility to explain diffraction phenomena, thereby definitely establishing the wave character of radiation. After Duane's demonstration that diffraction may be explained consistently based on particle properties of photons, Bohr preferred to ignore this possibility. The discovery of the Compton Effect did not change Bohr's attitude, instead the "BKS theory" introduced additional "virtual" fields and claimed energy and momentum conservation to be valid only on average, but violated for individual transitions. When the Bothe-Geiger experiment demonstrated support for the existence of photons and the validity of energy and momentum conservation for the individual transition, Bohr still was only partly convinced and did not fully renounce the wave concept. Schrödinger's wave functions were a welcome tool to justify a new principle, the Complementarity principle. Wave and particle properties—although mutually contradictory—should be complementary, since—as Bohr claimed—would never show up in the same type of observation. Interference of waves and diffraction should still be due to the wave properties; other observations demonstrating particle properties, such as the Compton effect, required different experimental arrangements, complementary to wave observations.

As already discussed in Sect. 7.1, Bohr's Complementarity principle and wave-particle duality is misleading. Experimental demonstrations of diffraction pattern building up from single particle processes are obtained routinely in neutron scattering. The appendix will contain further details.

# Chapter 8
# Early Opposition to the Copenhagen Interpretation

**Abstract** This chapter describes the decided opposition of Einstein and Schrödinger to the Copenhagen interpretation. Einstein's understanding of Quantum Theory is described in detail based on his own writings. According to Einstein the "EPR paper" (Einstein, Podolsky, Rosen, 1935) was written by Podolsky and does not properly contain Einstein's own stance.

**Keywords** Complementarity principle · Disturbances by measurements · Criticism by Schrödinger and Einstein · Heisenberg-Bohr tranquilizing philosophy · Einstein-Podolsky-Rosen ("EPR") paper · Incompleteness · "Ensemble of systems" · Born's scattering paper · Separability · Ensemble narrowing (subsemble) · Unified field theory

During the decades following the discovery of the new quantum laws the Copenhagen interpretation became the generally accepted version to "explain" Quantum Theory. Before 1925 Bohr had acquired the prestige of highest quantum authority; when the breakthrough in 1925 occurred, Bohr quickly nostrified the content of the new theory. He tailored both matrix mechanics and wave mechanics to suit his own preceding concepts. In Bohr's view, stationary states and discrete spectra had been and should remain to be the central elements of atomic physics. Heisenberg obliged; he was happy that his own reinterpretation paper was recognized as the essential novelty, providing support for Bohr's ideas. The contributions and radical changes contained in the papers of Born-Jordan—if recognized at all—were given second roles. During late 1925 and early 1926 Born was away in the United States and when he returned, Schrödinger's wave representation pulled attention away from the unfamiliar matrix representation. Heisenberg joined Bohr in Copenhagen, and public attention was centered on the controversy between Copenhagen and Schrödinger. Although Born's statistical interpretation of wave functions awarded only mathematical—not physical—significance to Schrödinger's waves, equivalence between matrix and wave representation enabled Bohr to adopt waves as providing proof of dual properties. Quantum uncertainties were claimed to result from unavoidable disturbances caused by measurements. Bohr's prestige, supported by Heisenberg and Pauli, prevailed in general recognition.

© The Author(s) 2017

H. Capellmann, *The Development of Elementary Quantum Theory*,
SpringerBriefs in History of Science and Technology,
DOI 10.1007/978-3-319-61884-5_8

Born did not hesitate to use Schrödinger's position representation, where its use provided mathematical advantages. Later he adopted the expression 'Complementarity'—albeit with meaning different from Bohr's intention—to indicate the possibility of the two mathematical representations. But he retained his convictions concerning the physical content of Quantum Theory. Born did not confront Bohr directly, he shied away from open controversy and avoided public confrontations, hoping that eventually his own contributions would be recognized at their correct value. The attribution of the Nobel Prize for the year 1932 to Heisenberg alone was a severe disappointment.

Very quickly after Bohr's proposal of the Complementarity principle, strong criticism was raised by Schrödinger and Einstein, who shared common convictions concerning what a valid quantum theory should rely upon. They both attributed "part of the truth" only to the Göttingen version of quantum theory. In particular, they refused to abandon the space-time continuum; differential equations should govern the behavior of the quantum world. Most importantly, they agreed that scientific laws must necessarily contain a consistent logical structure without internal contradictions; mutually contradictory notions have no place in science. Even if they disagreed on the reality and physical significance to be attributed to wave functions, both Schrödinger and Einstein refused to hide lack of logical structure under some "principle".

During the spring of 1928 several letters were exchanged between Schrödinger and Einstein, documenting their opposition to the Copenhagen interpretation. Bohr had sent his article introducing the Complementarity principle (Naturwissenschaften 15, 245–257, 1928) to Schrödinger, who quickly understood that the Copenhagen interpretation contained internal contradictions. Schrödinger's letter to Bohr of 5 May 1928 points out, that Bohr's concept of stationary states and discrete spectra is incompatible with the uncertainty relation. Two examples are discussed, where the internal contradictions are evident. The first example refers to the uncertainty relation between action variable $J$ and angle variable $\omega$. Periodicity with respect to angle over a period 1 leads to a maximum uncertainty $\Delta \omega = 1$ and a minimum uncertainty of the action variable $\Delta J \approx h$. Not only a perfectly sharp value of $J$ is ruled out, but also the distance between supposedly two distinct values of $J$ is of order the minimum uncertainty required. The second example discusses momentum quantization of a free particle inside a finite volume $V$. The maximum position uncertainty is limited by the volume, leading to a minimum momentum uncertainty, which again is of order the difference in momentum between neighboring values. Furthermore Schrödinger criticized Bohr for his use of notions invoking unobservable processes. "It seems to imperatively require the introduction of new notions, which do not have these limitations. What is unobservable in principle should not be part of our terminology."

Bohr was not to be impressed by such "formal" arguments; his reply to Schrödinger of 23 May 1928 explained all problems by "the recognition of the inevitable trait of complementarity, which is exposed in the analysis of the observational concept, and which, in many respects is reminiscent of the recognition of the general theory of relativity". If there seemed to exist contradictions, then that proved the necessity for complementary notions!?

Bohr had asked Schrödinger to inform Einstein of his views; Schrödinger sent Einstein a copy of his own letter to Bohr as well. Einstein's reply to Schrödinger of 31 May 1928 expresses his agreement with Schrödinger together with sarcastic irony concerning the Copenhagen interpretation, in particular Bohr's complementarity and dual properties:

> "The Heisenberg-Bohr tranquilizing philosophy—or religion—is so delicately concocted that it provides a gentle pillow for the believer from which he cannot be aroused that easily. So let him lie there. But this religion has so damned little effect on me that despite everything I say: Not: E and $\nu$; but rather: E or $\nu$; and to be precise: not $\nu$, but E (since it has ultimately reality.)"

Photons just as other particles have energies and momenta, not frequencies and wavelengths. Advocating mutually exclusive properties simultaneously did not only violate classical logic, it contradicted any logic.

## 8.1  Einstein's Understanding of Quantum Theory

The "EPR paper" by Einstein, Boris Podolsky, and Nathan Rosen ("Can Quantum-Mechanical Description of Physical Reality be Considered Complete?", Phys. Rev. 47, 777–881, 1935) is one of the most famous publications on Quantum Theory. Considerably less known is the fact, that Einstein quickly distanced himself from parts of the EPR paper. On 19 June 1935 he sent a letter to Schrödinger, stating that Podolsky wrote the paper, but that his own (Einstein's) views were not represented properly. "The principle message was buried by punditry"("Die Hauptsache ist sozusagen durch Gelehrsamkeit verschüttet"). The Institute of Advanced Studies Letter of Fall 2013 contains an article by Kelly Devine Thomas "The Advent and Fallout of EPR", which gives more details about the origin of the EPR paper. The EPR paper appeared in print on May 8, 1935. Four days before, on May 4, the New York Times contained an article: "Einstein Attacks Quantum Theory". Podolsky was the originator, without informing Einstein. On May 7, the New York Times published Einstein's rebuttal, expressing his anger. According to Asher Peres, whose Ph.D. thesis adviser was Rosen, Einstein never spoke to Podolsky afterwards.

If we want to know Einstein's reasoning, we have to look at his own writings, not at Podolsky's "Gelehrsamkeit". Einstein addressed the matter of the EPR paper in his publication "Physik und Realität" as part of a broad review (Journal of the Franklin Institute Vol. 221, 313–347, March 1936. A translation by Jean Picard from German into English follows on pp 349–382).

Einstein's own publication is clear and straightforward: Born's statistical and discontinuous interpretation of Quantum Theory is recognized to properly describe the experimentally observed phenomena. The physical content of Schrödinger's continuous representation is reduced to the original form of the quantum laws derived by

Born and Jordan. But Einstein does not consider Quantum Theory to contain the final verdict of all what is possible; Quantum Theory is termed to be incomplete: "Der Unvollständigkeit der Darstellung entspricht aber notwendig der statistische Charakter (Unvollständigkeit) der Gesetzlichkeit". ("The incompleteness of the representation, however, necessarily corresponds to the statistical character (incompleteness) of the laws". (my translation)). Einstein is not satisfied by this statistical (and therefore—in his opinion—incomplete) character; he calls for a deeper and complete theory, **"a uniform basis for physics"**, based on differential equations. This field theory should restore the space-time continuum and eliminate the statistical feature, enabling a deterministic description of the individual system.

Two examples are used to show that the Schrödinger equation does not represent the physical state of an individual physical system completely. The first example treats a single quantum system, the normalized eigenfunctions are $\psi_i$ with eigenvalues $E_i$; the spectrum may be partly discrete and continuous. Let us be more specific and take the system to be the hydrogen atom; generalizations to other systems are straightforward. Let the initial state be represented by $\psi_1$. Now a weak external perturbation, e.g. an external potential, is applied for a finite time. According to the Schrödinger equation, the initial $\psi$ function is transformed into $\psi(t) = \sum c_i(t)\psi_i$. Since the external force is required to be small, $|c_1|$ is close to one, the other non-vanishing $|c_i|$ are small. Einstein concludes:

> "It follows that our function $\psi$ does not in any way describe a homogeneous condition of the body [in our special case the atom], but represents rather a statistical description in which the $c_i$ represent probabilities.....the Born statistical interpretation of the quantum theory is the only possible one. The $\psi$ function does not in any way describe a condition which could be that of a single system; it relates rather to many systems, to "an ensemble of systems" in the sense of statistical mechanics." (The quotation marks for "an ensemble of systems" are Einstein's.)

Einstein concludes that the physical state of the atom does not contain fractions of vastly different energies; the Schrödinger equation does not describe the continuous change of the physical state with time; only transition probabilities to different physical states may be obtained. The $c_i$ are related to probabilities for the occurrence of a transition.[1] The difference to classical expectations is pointed out:

"While in classical mechanics, such additions [i.e. the small external forces] can produce only correspondingly small alterations of the system, in the quantum mechanics they produce alterations of any magnitude however large, but with correspondingly small probability".

The reason for the inherent statistical nature is not due to some disturbance introduced by measurements, as Heisenberg and Bohr argued, but is required by the structure of the laws themselves. The Schrödinger equation obtains physical sense only if combined with Born's statistical interpretation. Einstein explicitly attacks the Copenhagen doctrine of disturbance:

---

[1] The probabilities may be calculated in various ways; "time"-dependent perturbation theory is the standard method used. The appendix gives another version to obtain the same result.

"If, except for certain special cases, the $\psi$ function furnishes only *statistical* data concerning measurable magnitudes, the reason lies not only in the fact that the *operation of measuring* introduces unknown elements, which can be grasped only statistically, but because of the very fact that the $\psi$-function does not, in any sense, describe the condition of *one* single system. The Schrödinger equation determines the time variations which are experienced by the ensemble of systems". (The emphasis in italics is Einstein's.)

Let us summarize Einstein's conclusions: The function $\psi(t) = \sum c_i(t)\psi_i$ does not describe the continuous time evolution of the **physical state** of one particular system; the **physical state** changes discontinuously and statistically. An ensemble, i.e. a large number of identical and independent systems (for our special case independent atoms), will, after the application of the small perturbation for a finite time, have evolved into the following physical situation: Most systems will be unaltered ($|c_1|$ is close to one), whereas small numbers have undergone transitions into other states with energies $E_i$, as indicated by the small values of the other $|c_i|$. The distribution function is obtained from the ensemble average. Einstein restates the essential difference between classical expectations and Quantum Theory. Classically, very weak external forces may change the physical state only by accordingly small amounts; however

"quantum mechanics can easily account for the fact that weak disturbing forces are able to produce alterations of any magnitude in the physical condition of a system. Such disturbing forces produce, indeed, only correspondingly small alterations of the statistical density in the ensemble of systems, and hence only infinitely weak alterations of the $\psi$ functions."

Einstein has adopted Born's interpretation: The laws of Quantum Theory are inherently statistical, describing discontinuous transitions.

The path, which led Einstein to this conclusion, is remarkably different from Born's. Already very early on, Born became convinced that the space-time continuum lost its meaning on atomic scales, as Born's letter to Pauli of December 1919 testifies. Quantum transitions should occur in discontinuous and statistical steps. Einstein's own papers had led Born to this conviction. But Einstein's reaction to Born's concept (letter of 27 January 1920) was negative: "I do not believe that the continuum has to be abandoned to solve the quantum problem. For in analogy one might argue that general relativity might be wrested from the elimination of the coordinate system." Einstein will remain consistent throughout, the ultimate aim will be the restoration of the continuum and relativity will remain to be a central argument. Born pursued his program. Quantization of action required discontinuous transitions, discontinuities required statistical behavior. When the quantum laws were finally obtained by Born and Jordan, their purely algebraic form reflected Born's longterm program. When Schrödinger's version of quantum mechanics appeared, Einstein's first reaction was very positive; a continuous field $\psi$ changing continuously in time corresponded to his own expectations. But upon closer inspection Einstein's enthusiasm disappeared; the very structure of the Schrödinger equations and their solutions—Einstein concluded—enforced Born's physical interpretation.

The second example discussed in Einstein's review treats the consequences of interaction between two quantum systems. In the EPR paper Podolsky had reached incorrect conclusions concerning this problem; his central argument relied on his incorrect interpretation of the wave-function describing a composite system of two spatially separated systems (A and B), which had previously interacted. Discussing different measurements on system A, Podolsky claimed—incorrectly—that the unmeasured system B may have precise values of noncommuting observables simultaneously.[2] Einstein clearly rejects Podolsky's incorrect arguments; instead he refers to his ensemble interpretation: "Such an interpretation eliminates also the paradox recently demonstrated by myself and two collaborators".

Explicitly the following situation is discussed. The initial state is taken to consist of two quantum systems spatially separated and noninteracting. Then the two quantum systems interact for a limited time; in the final state the two systems are again spatially separated and noninteracting. Born's scattering paper (Z. Phys. 37, 867, 1926), which introduced the statistical interpretation of $\psi$, is directly related to Einstein's arguments; let us take the specifications of Born's paper. The initial $\Psi$ function, $\Psi_1 = \psi_i^p \psi_j^a$, represents a noninteracting state of a particle (p) and an atom (a) far apart from each other. The initial particle state of finite momentum is represented by $\psi_{i,}^p$, the atom by $\psi_j^a$. A collision between particle and atom leads to a scattering process from the initial state to a final state caused by the particle-atom interaction; the final state is taken to consist of particle and atom again far apart and noninteracting.

The Schrödinger equation describes the "time"-evolution from the initial $\Psi$ function—assumed to be given - into a final function $\Psi_f(t) = \sum_{mn} C_{ij,mn}(t) \psi_m^p \psi_n^a$. The $\psi_m^p$ are taken to be a complete system of particle functions and the $\psi_n^a$ a complete system of atom functions. Einstein requires that in the final state no measurement on one of the systems may influence the physical properties of its previous partner, now far away and noninteracting. At first sight this seems to be in conflict with the Schrödinger equation. The $\Psi_f$ obtained from the Schrödinger equation, $\Psi_f(t) = \sum_{mn} C_{ij,mn}(t) \psi_m^p \psi_n^a$, associates a specific final state $\psi_k^p$ of the particle with a specific, i.e. $k$-dependent, $\psi_{k,f}(t) = \sum_n C_{ij,kn}(t) \psi_n^a$. According to the separability argument, different measurements performed on the particle in the final state—corresponding to different quantum numbers $k$—cannot influence the physical properties of the atom far away and noninteracting. Einstein concludes: Therefore the $\psi_{k,f}$ cannot represent different **physical states** of the atom, and $\Psi_f(t) = \sum_{mn} C_{ij,mn}(t) \psi_m^p \psi_n^a$ does not describe the final **physical state** of the combined particle-atom system. $\Psi_f = \sum_{mn} C_{ij,mn} \psi_m^p \psi_n^a$ allows statistical conclusions only about the outcome of possible transitions. Finite $C_{ij,mn}$ indicate finite probabil-

[2]On the other hand, Podolsky shows that the wave function of a single system does not allow this system to have precise values of noncommuting observables simultaneously. Via some weird chain of arguments Podolsky then reaches the conclusion that the wave-function of the single system cannot be complete. One has to read the paper at least five times to follow the chain of arguments; the whole thing is so un-Einstein like if compared to the clarity of Einstein's own publications, that it is hard to believe that Podolsky had Einstein's approval to publish this version.

ities $P(ij, mn)^3$ for transitions from initial states represented by $\psi_i^p \psi_j^a$ into physical states represented by $\psi_m^p \psi_n^a$. Einstein marks his disagreement with supposedly paradoxical features of Quantum Theory, such as spurious instantaneous influences of measurements on spatially separated systems.

Explicitly Einstein insists that "the coordination of the $\psi$ function to a ensemble of systems eliminates every difficulty".

To clarify the ensemble concept, Einstein mentions that measurements performed on one of the subsystems lead to an ensemble narrowing (subensemble), and that different types of measurements lead to different subensembles. Einstein's remarks are rather short, a few further specifications may be helpful. The example used above consisting of particle (p) and atom (a) may be mapped on Einstein's by attributing internal degrees of freedom to both, atom and particle. The particle may have spin, the atom may have magnetic moment, nuclear spin, different internal energies. The particle indices $I$ and $M$ now stand for combinations of quantum numbers necessary to fully characterize particle states, similarly $J$ and $N$ for the atom states. Apart from this replacement, the notation used above is retained. The previous conclusions remain valid; the coefficient $C_{IJ,MN}$ indicates that there exists a transition probability $P(I, J, M, N)$ from initial state represented by $\Psi_1 = \psi_I^p \psi_J^a$ to a final state represented by $\Psi_f = \psi_M^p \psi_N^a$.

Consider an ensemble of identical initial systems represented by $\Psi_1 = \psi_I^p \psi_J^a$.

The total number of scattering processes is proportional to the probability summed over all possible final states $\sum_{M,N} P(I, J, M, N)$. Now let the experimental setup select particles of final momentum $\mathbf{k}$. Let $M_\mathbf{k}$ stand for combinations of particle quantum numbers for fixed final momentum $\mathbf{k}$. The corresponding transition probability is obtained from the average over a narrower ensemble, i.e. $\sum_{M_\mathbf{k}N} P(IJ, M_\mathbf{k}N)$. A further reduction of the ensemble is obtained, if the experiment selects scattered particles with spins in some chosen direction $\Theta$ only. The particles are now fully characterized by $M_{\mathbf{k}\Theta}$; their number is proportional to the average over the subensemble of possible final atomic states: $\sum_N P(IJ, M_{\mathbf{k}\Theta}N)$.

Different final momenta $\mathbf{k}$ and different directions $\Theta$ result in different subensembles.

Einstein recognizes the logical consistency of Born's concept, i.e. discontinuous and statistical quantum transitions as fundamental elements of Quantum Theory. Nevertheless he does not accept it as "end of story"; he specifies: "To believe this is logically possible without contradiction; but, it is so very contrary to my scientific instinct that I cannot forgo the search for a more complete conception." The space-time continuum remains to be the foundation for Einstein's understanding. He does not accept the discontinuous and statistical theory of Born-Jordan as a **final** theory, since then "we must also give up, by principle, the space-time continuum....At the present time, however, such a program looks like an attempt to breathe in empty space." His aim is a unified field theory, "a uniform basis for physics", which should

---

[3]For weak interactions the probabilities are given in Born's scattering paper. "Time"-dependent perturbation theory may also be used; equivalent results are obtained using the methods of the appendix.

eventually eliminate the statistical character and reestablish continuity in space and time.

The acceptance of Born's statistical interpretation of Quantum Theory is confirmed in Einstein's letter to Born of 18 March 1948. Concerning "causality of the observable (Kausalität des Beobachtbaren)" Einstein concedes: "I do know that causality does not exists with respect to the observable; I consider this cognition to be final (Ich halte diese Erkenntnis für endgültig)." Nevertheless Einstein insists: "But in my opinion one should not conclude that the theory should rely on statistical principles.... One cannot consider the quantum mechanical description to be a complete representation of what is physically real (des Physikalisch-Realen)."

Einstein repeated and expanded his understanding in "Albert Einstein: Philosopher-Scientist"(ed. P. A. Schilpp; Cambridge Univ. Press, 1949).

> "Within the framework of statistical quantum theory there is no such thing as a complete description of the individual system.....I am convinced that everyone who will take the trouble to carry through such reflections conscientiously will find himself finally driven to this interpretation of quantum-theoretical description (the $\psi$-function is to be understood as the description not of a single system but of an ensemble of systems)".

The "Heisenberg-Bohr tranquilizing philosophy" is rejected; statistical elements are not due to disturbances introduced by measurements; the disapproval of Bohr's complementarity is expressed in a slightly more polite manner, but leaving no doubt as far as his opinion is concerned:

> "...it must seem a mistake to permit the theoretical description to be directly dependent upon acts of empirical assertions, as it seems to me to be intended [for example] in Bohr's principle of complementarity, the sharp formulation of which, moreover, I have been unable to achieve despite much effort which I have expended on it".

A deeper and more complete theory should be the aim; to indicate its relation to Quantum Theory; Einstein refers to examples taken from classical physics. Phenomenological thermodynamics is recognized as a theory never to be overturned within the range of its applicability. The important specification is "within the range of its applicability", which is macroscopic physics. On a more elementary scale, Boltzmann's kinetic theory and statistical mechanics take over. Similarly, as far as the detailed description of individual processes is concerned, statistical mechanics is to be replaced by deterministic classical mechanics. Transferring this concept of "scale separation", Einstein requires that the future complete theory should contain and be compatible with Quantum Theory as a limiting case. But just as statistical mechanics cannot be derived from thermodynamics and Newtonian mechanics cannot be derived from statistical mechanics, Quantum Theory—in Einstein's opinion—should not constitute the basis for the derivation of this future complete theory.

It is incorrect to portray Einstein as simply opposing Quantum Theory, Einstein explicitly accepts the validity of Born's statistical and discontinuous interpretation of Quantum Theory. He rejects the Bohr-Heisenberg interpretation and he rejects any physical significance of wave-functions other than mathematical tools to provide probabilities. But there remained a profound disagreement between Einstein and Born concerning the ultimate status of Quantum Theory. Born was convinced that the space-time continuum had to be abandoned on the quantum scale once and for all; there existed no road back to continuity and determinism. Born was inspired by Einstein's Relativity Theory, which required a fundamental redefinition of space and time. But it retained the continuum; Born's "Quantenmechanik" required an equally dramatic change. On the elementary quantum scale the continuum itself has lost its meaning. That is what Einstein did not want to accept, he was convinced that there should be a unified continuum description of nature. The complete field theory— envisaged by Einstein—should contain and be compatible not only with Quantum Theory, but also Relativity Theory, within their respective domains of applicability. That was Einstein's goal; there is high probability that it is unattainable. Einstein himself did not succeed, but he confirmed:

"Continuous functions in the four-dimensional [continuum] as basic concepts of the theory. Rigid adherence to this program can rightfully be asserted of me".

# Chapter 9
# Orthodox Portrayals of the Development of Quantum Mechanics, Comparison and Differences

**Abstract** This chapter contains discussions of orthodox accounts of the development of Quantum Mechanics. The classical books by Max Jammer (1966) and B. L. van der Waerden (1967) represent the Copenhagen point of view; the multi-volume work of Jagdish Mehra and Helmut Rechenberg (1982) is similar in general conclusion. The essential differences to the conclusions reached in the present account are illustrated.

**Keywords** Copenhagen view · Virtual oscillators · Virtual radiation · Bohr-Kramers-Slater theory · Classical dispersion theory · Quantum vectors · Observability criterion · Basic law of quantum optics (Grundgesetz der Quantenoptik) · Measuring process · Quantum uncertainties · Interference of waves · Mutually exclusive notions · "Intermediate kind of reality" · Particle-wave duality

Roughly 4 decades after the crucial years 1925/1926 the first reviews of the historical development of Quantum Theory appeared. **Max Jammer's book "The Conceptual Development of Quantum Mechanics"** (Mc Graw Hill, 1966) contains the first comprehensive description covering the evolution from the old Quantum Theory up to the Copenhagen interpretation in 1927/28. In the following year **B. L. van der Waerden's book "Sources of Quantum Mechanics"** (North Holland, 1967) provided reprints of a selection of important publications from the period between 1916 and 1925 leading to the matrix representation of Quantum Mechanics; original German papers were translated into English "in order to make the sources available to all physicist and historians of Science." The introductory sections contain discussions and interpretations of the papers selected to be reprinted. Fifteen years later in 1982 the four volume series by **Jagdish Mehra and Helmut Rechenberg "The Historical Development of Quantum Theory"** (Springer, 1982) provided an extremely detailed account of the scientific activities which ultimately resulted in the formulation of matrix mechanics. These classical references promoted the Copenhagen view of the emergence of Quantum Mechanics. Heisenberg's influence is perceptible. Jammer states his gratitude to Heisenberg for "...discussing with him ... and reading the entire manuscript.." before publication. Van der Waerden had been Heisenberg's colleague at the University of Leipzig from 1931 to 1942, and Heisenberg's perspective dominates the selection of papers reprinted and discussed. Mehra,

© The Author(s) 2017
H. Capellmann, *The Development of Elementary Quantum Theory*,
SpringerBriefs in History of Science and Technology,
DOI 10.1007/978-3-319-61884-5_9

after completing his first degree in Physics in India, had worked with Heisenberg at the Max-Planck-Institute of Physics in Göttingen from 1952 until 1955; Rechenberg had been Heisenberg's Ph.D. student (Dr. in 1968).

The basic message contained in these orthodox portrayals is identical; Quantum Mechanics alone is addressed, described as independent of the quantum behavior of radiation. The old Bohr-Sommerfeld Quantum Theory, which rejected Einstein's Quantum Theory of radiation, is considered to contain the essential elements from which the new theory was built. Heisenberg's reinterpretation paper of July 1925 is argued to provide a calculational scheme for Bohr's physical concepts. Bohr's "Correspondence principle" is described to have provided the most important guideline. The Correspondence principle in its most general form had the meaning: "something in classical physics corresponds to something in quantum physics"; what these two "somethings" should be, however, varied in time and between different authors. The definition contained in Bohr's review article (Z. Phys. 13, 117–165, 1923) suggested that some harmonic oscillation process inside the atom should correspond to the frequency of continuous radiation emitted or absorbed. Since radiation was considered to be classical, it had to be emitted or absorbed by some oscillation process of the required frequency. This atomic oscillation processes became the virtual oscillators of the Bohr-Kramers-Slater theory. And these virtual oscillators generating continuous radiation are argued to have provided the connection from the old Quantum Theory directly to Heisenberg's treatment of the oscillation process in his reinterpretation paper of July 1925.

Jammer describes Bohr's Correspondence principle as "the most versatile and productive device for the further development of the older quantum theory....It was due to this principle that the older quantum theory became as complete as classical Physics." Nevertheless, concerning the conceptual understanding, Jammer concedes:"Quantum theory ...prior to 1925, was, from the methodical point of view, a lamentable hodgepodge of hypotheses, principles, theorems, and computational recipes rather than a logical consistent theory." Heisenberg's reinterpretation paper of July 1925 is argued to transform the "lamentable hodgepodge" into a logical foundation of Quantum Mechanics. The commutation relations and the quantum equations of motion derived by Born and Jordan in September 1925 (BJ-2) are considered to be mathematical extensions of Heisenberg's paper, just as the common paper of Born, Heisenberg, and Jordan (BHJ) of November 1925.

Similar to Jammer, van der Waerden presents the development of matrix mechanics to constitute the continuation from the old quantum theory; Heisenberg's reinterpretation paper is given the decisive role for the development of matrix mechanics. In the preface van der Waerden cites Born as the initiator for the format of the book: "The idea of collecting the most important papers on Quantum Mechanics in a Source Publication is due to Max Born, and he intended to include 15 papers written by himself, Jordan, Heisenberg, Dirac, and Pauli, and published during the years 1924–1926." The selection of papers actually reprinted in van der Waerden's book is substantially different; the Copenhagen perspective of Bohr-Heisenberg dominates. Only seven papers from Born's proposal were retained, the most important omission concerns the paper of Born-Jordan of June 1925 (BJ-1), which is not included among the papers reproduced and translated.

The book is divided into two parts; Part I "Towards Quantum Mechanics" contains 11 papers, covering the period from 1916 up to mid 1925. The emphasis is shifted towards the Copenhagen concept of oscillators interacting with continuous radiation and their connection to classical dispersion theory, which is supposed to have prepared the way towards Quantum Mechanics. Part II "The Birth of Quantum Mechanics" contains six papers starting with Heisenberg's reinterpretation paper of July 1925. Van der Waerden reports that Heisenberg confirmed the principle idea:"something in the atom must vibrate with the right frequency" to produce the continuous radiation during the transition process. The treatment of the oscillator concept provided the connection between the old Quantum Theory and Heisenberg's paper. Van der Waerden's discussion of the Born-Jordan publication "Zur Quanten-mechanik" (BJ-2) of September 1925 does mention that quantization of radiation is contained in BJ-2, but due to the omission of the Born-Jordan paper (BJ-1) of June 1925 the connection between Einstein's photons—eliminating the continuous behavior of radiation—and the ideas of Born and Jordan—radically different from the old Quantum Theory and eliminating the virtual oscillators—are absent.

Chapter 5 of the present book puts Heisenberg's paper into the context of the preceding work of Born and Jordan and the developments following afterwards. In this context, it is Heisenberg's paper—although an important contribution—which appears more to be a mathematical contribution providing a multiplicational scheme for the quantum vectors, introduced previously by Born and Jordan. The June 1925 paper (BJ-1) defines the path to follow: Einstein's "Quantenoptik" is connected to Born's "Quantenmechanik" via quantum vectors (= matrix elements) representing radiative transitions for emission and absorption of light quanta; transition probabilities are proportional to the absolute squares of the quantum vectors. Heisenberg contributes the multiplication scheme for the quantum vectors in July. But the decisive breakthrough is achieved in September 1925 (BJ-2) again by Born and Jordan, who discover the basic laws of Quantum Theory: Commutation relations, quantum equations of motion, and a quantization scheme for the radiation field.

The very detailed account by Mehra and Rechenberg again attributes to Heisenberg the central role for the development of the Göttingen version of Quantum Mechanics. But in contrast to the books by Jammer and van der Waerden, the account by Mehra and Rechenberg contains extensive descriptions of preceding contributions by Born and Jordan, which are important for Heisenberg's paper. It is mentioned that, upon Heisenberg's arrival in Göttingen to start his new position in Göttingen in October 1923, Born had defined the program "discretization of atomic physics"; Born's elaborations on this program in his lectures during the winter of 1923/24 are equally mentioned. The paper by Born "Über Quantenmechanik"of 1924 and in particular the paper by Born and Jordan of June 1925 (BJ-1) are discussed. Credit is given to Born and Jordan for applying the principle of retaining only observable quantities, explicitly quoting their paper: 'A fundamental axiom of large range and fruitfulness states that the true laws of nature involve only such quantities as can be observed and determined in principle'. Furthermore, the comparison made by Born and Jordan to

Einstein's use of the observability criterion to derive Relativity Theory is mentioned as well. Mehra and Rechenberg continue: "All this had taken place before Heisenberg himself ever invoked the principle of observability of physical quantities in the actual construction of a theory. One cannot find any reference to it in any of his earlier papers or scientific correspondence." They conclude that Heisenberg did not rely on the observability principle to derive his results concerning the enharmonic oscillator, when he spent part of June 1925 in Helgoland. After his return from Helgoland to Göttingen, he again took part in discussions with Born and Jordan. Only these belated discussions—Mehra and Rechenberg point out—stimulated Heisenberg to include this criterion in his own paper.

Nevertheless Mehra and Rechenberg do not expose the essential results of the paper BJ-1 by Born and Jordan, i.e. the connection of Einstein's Quantum Optics and Born's Quantenmechanik. Einstein's papers of 1916/17 are discussed in chapter V of volume I; Einstein's insistence on the necessarily quantal character of radiation is explicitly mentioned. This quantal character is fully retained by Born and Jordan; the paper BJ-1 develops quantum mechanical perturbation results and describes energy transfer between matter and radiation in quantized entities. According to Mehra-Rechenberg, however, the June 1925 paper of Born-Jordan was able "to verify Einstein's derivation of Planck's radiation law, without invoking the concept of light-quanta."

A consistent Quantum Theory cannot combine **discontinuous mechanical behavior** with emission and absorption of **continuous radiation**. Radiation has to be quantized as well. The paper BJ-1 by Born and Jordan relates Quantum Mechanics—i.e. discontinuous and statistical quantum transitions—to Einstein's "basic law of Quantum Optics (Grundgestz der Quantenoptik)". The term "Quantenoptik" is introduced by Born and Jordan in this paper. Concerning the relations between Heisenberg's paper of July 1925 and the two papers by Born and Jordan of June (BJ-1) and September (BJ-2), remarks made about the books by Jammer and van der Waerden apply equally to the account of Mehra-Rechenberg: The path leading from BJ-1 to Heisenberg's paper and finally to the breakthrough contained in BJ-2 does not become apparent. No connection is established between Born and Jordan's "quantum vectors" (later to be called matrix elements) in BJ-1 and their adoption by Heisenberg as "komplexe Vektoren" to represent quantum theoretical amplitudes. The logical connection is missing between the physical concept contained in Born's Quantenmechanik of 1924 and the fundamental equations of Quantum Theory established by Born and Jordan in September 1925.

Jammer's book also covers the further developments up to 1928, which are described following the line of thought of Heisenberg and Bohr. Heisenberg had left Göttingen in the spring of 1926 and started his new regular position at Bohr's institute on May 1st 1926. Bohr interpreted the new developments as confirmation of his own preceding ideas. He considered Heisenberg's reinterpretation paper to constitute the confirmation of his own virtual oscillators. Jammer reports: "To him, of course, there was nothing alien in it. On the contrary, he declared. 'The whole apparatus of quantum mechanics can be regarded as a precise formulation of the tendencies embodied in the correspondence principle'." Heisenberg's understanding was similar, he considered his own paper to provide a calculational method for oscillators generating continuous radiation. For Bohr, too, that was the logical way to fit his own thinking; for him, quantized energies of stationary states and the frequency condition had been and remained to be the decisive elements. The frequency condition required an underlying oscillation process emitting continuous radiation; Heisenberg's oscillator amplitudes provided the intensities. A full understanding of the fundamental quantum laws derived by Born and Jordan had not reached Copenhagen even after Heisenberg's arrival.

The search for a deeper understanding leads Heisenberg towards "Anschaulichkeit" of commutation relations and disturbance induced indeterminacy of quantum uncertainties. Jammer's description of the uncertainty relations accepts Heisenberg's erroneous arguments as valid. The measuring process is argued to necessary alter the state of the system to be measured. Quantum uncertainties are not considered to be constitutive elements of quantum Physics, and Quantum Theory is not considered to be essentially a statistical theory. Statistical behavior is blamed on the observational process. Jammer writes "Heisenberg drew a conclusion of far-reaching philosophical implications at the end of his paper", quoting Heisenberg (Jammer's translation):

"We have not assumed that the quantum theory, unlike classical physics, is essentially a statistical theory in the sense that from exact data only statistical conclusions can be inferred.... in the strong formulation of the causal law If we know exactly the present, we can predict the future' it is not the conclusion but rather the premise which is false. We cannot know, as a matter of principle, the present in all its details."

The rejection of Einstein's quanta of radiation had been justified by Bohr, because—in his opinion—diffraction phenomena could only be explained by interference of waves. As mentioned in Chap. 7, Duane, in 1923, disputed this point of view and gave an explanation consistent with Einstein's particle concept of light quanta. He suggested that diffraction is caused by quantized momentum transfers in particle scattering processes. Jammer recognizes Duane's argument, but he restricts its applicability: "As Duane's work and Compton's elaborations of it showed, Fraunhofer diffraction phenomena could be accounted for in terms of the quantum-corpuscular theory of light. All attempts, however, to obtain similar results for finite diffraction systems or Fresnel diffraction phenomena proved unsuccessful."

These arguments are used to justify the introduction of dual properties, particle and wave-like, which Bohr postulated when he introduced his Complementarity principle. Jammer acknowledges that the interaction of light and matter suggests particle properties of light; interference phenomena such as diffraction, however, are supposed to require wave character. Attempts to explain the latter in the particle picture should necessarily attribute a frequency to the photon, and thereby also a wavelength. Wavelengths, however, could—in Jammer's opinion—only be determined by diffraction invoking wave properties: "In other words, a hypothesis supported by incontestable experimental evidence, became physically significant only by the use of its own negation....", leading to Jammer's conclusion: "...it became clear that the interpretation of optical phenomena required....the use of mutually exclusive notions."

Schrödinger's wave mechanics and wave functions became essential elements for Bohr's Complementarity principle. In support of Bohr, Jammer attributes a kind of "intermediate reality" to Schrödinger's wave functions: "....the ontological status of $\psi$ had to be considered as something intermediate....precisely that "intermediate kind of reality" which, as Heisenberg had emphasized, transpired in the work of Bohr, Kramers, and Slater in 1924." Jammer joins the architects of the Copenhagen interpretation and attributes new esteem to the very doubtful value of the BKS theory. He continues: "Now it will be understood that Born's probabilistic interpretation was indeed influenced by the Bohr-Kramers-Slater-conception of the virtual radiation field". Bohr's complementarity principle completes what Jammer considered "...the conceptual situation as brought about by the establishment of the so-called Copenhagen interpretation ..... de fact the only existing fully articulated consistent scheme of conceptions that brings into order an otherwise chaotic cluster of facts and makes it comprehensible".

Chapters 5–7 of this book lead to different conclusions. Comprehension cannot be achieved by adding mutually exclusive notions to classical concepts. Particle-wave duality is misleading; so-called "interference"—effects result from particle scattering. The scattering probabilities may be represented mathematically in various ways, but there is no justification for a physical wave property of single particles. And quantum uncertainties are not due to disturbances introduced by the observational process but are constitutive elements of quantum Physics.

# Chapter 10
# Later Criticism of the Copenhagen Interpretation

**Abstract** This chapter describes later opposition to the Copenhagen interpretation. The logical consistency of the Copenhagen interpretation—in particular Bohr's "philosophical" pronouncements—are viewed with growing skepticism by historians and philosophers of science.

**Keywords** "Quantum Fact and Fiction" (Alfred Landé) · Duane's quantum rule · "The Shaky Game" (Arthur Fine) · Heisenberg-Bohr tranquilizing philosophy · EPR paper · Einstein's ensemble interpretation · Hidden variables · Bell's theorem · Bohr-Einstein controversy · Philosophical rhetoric

Although the Copenhagen interpretation dominated, decided opposition developed as well. E.g. in 1965 **Alfred Landé** attacked Bohr's mysticism in the paper **"Quantum Fact and Fiction"** (Am, J, Phys. 33, 123, 1925). "Fiction" referred to dual nature of matter, "Fact" to Duane's quantum rule for momentum exchange introduced in 1923 (described extensively in Chap. 7). In addition to Einstein and Schrödinger, Landé could cite growing criticism among philosophers of science like **Henry Margenau and Karl Popper**, who attacked the Copenhagen doctrine (Popper: *"the ultimate betrayal of Galilean science"*). Four decades after the crucial year 1925, the situation as described by Landé was truly strange. A large fraction of the scientific community accepted the Copenhagen doctrine that one should not even try to understand quantum behavior. Instead the supreme authority's verdict was taken as final; one has to "refine the language of physics" and simply accept irrational, mutually exclusive notions. Although Bohr's pronouncements were notoriously vague and difficult to understand, that was taken to be indicative of Bohr's deeper insight. Others, however, dared to come to the conclusion that Bohr himself did not understand what Bohr was saying.

At the end of his paper, Landé makes a comment concerning the different interests of 'physicists busy with research at the frontiers of science' on one side and 'students who ask about the "why" of the baffling techniques' on the other. The physicist-researcher might not be interested in the general roots of the theory he is using as a tool to perform calculations; students, however, might want to understand the

© The Author(s) 2017
H. Capellmann, *The Development of Elementary Quantum Theory*,
SpringerBriefs in History of Science and Technology,
DOI 10.1007/978-3-319-61884-5_10

basic principles themselves, without simply having to accept the verdict of higher authority.

Landé's paper concludes with "... after an age of ambiguity dominated by the Bohr-Heisenberg duality as a "principle", the quantum philosophy is open for an agonizing reappraisal."

In 1986 **Arthur Fine** voiced further criticism of the Copenhagen doctrine in his book **"The Shaky Game, Einstein Realism and the Quantum Theory"** (The University of Chicago Press, 1986, 2nd ed. 1996). Fine presents a detailed analysis of Einstein's attitude towards Quantum Theory and of his reaction to the establishment of the Copenhagen interpretation. The defense of Einstein starts with the legend that the young and revolutionary Einstein changed into an older, conservative Einstein, who was unable to accept the new developments of the quantum revolution of 1925–1927. At close view, this prejudice is truly strange, in particular if contrasted to Bohr's "innovative" concepts culminating in the Complementarity principle. Fine briefly recalls Einstein's radical break with continuum physics, when he introduced the photon in 1905, which Bohr consistently contested for decades, even saving the wave aspect in complementarity. It should be added that in 1924/25 Einstein showed extraordinary audacity to fill Bose's statistics with its physical meaning of indistinguishability, to apply this new principle to atomic gases, to predict Bose-Einstein condensation, and to suggest interference-type phenomena for particles with finite mass. Bohr tried to save classical concepts in his Complementarity principle, rejected by Einstein as part of the "Heisenberg-Bohr tranquilizing philosophy". It was Bohr, who maintained that classical concepts must be maintained to understand Quantum Theory, whereas Einstein, even in his criticism of the new theory, advocated to go further towards a new type of unified field theory. Fine concludes: "Thus the tale of Einstein grown conservative in his later years is here seen to embody a truth dramatically reversed. For it is Bohr who emerges the conservative, unwilling (or unable?) to contemplate the overthrow of the system of classical concepts".

Based on Einstein's correspondence Fine demonstrates that Einstein's basic attitude towards the emerging quantum theory developed rather quickly; already in early 1927 he had reached the conclusion, that both matrix and wave mechanics could yield statistical results only and were incapable of describing the behavior of individual systems completely. Well before the EPR paper Einstein had accepted the validity of the uncertainty relations. The issues raised in the EPR paper are discussed in detail. Considering the fact that Podolsky wrote this paper and given the ambiguous, vague, and unclear structure of the EPR paper, Fine suggests that Einstein may not have seen the draft before submission.[1]

Fine undertakes the difficult task to extract Einstein's understanding from under the EPR erudition; furthermore Einstein's own writings are analyzed in detail. His correspondence with Schrödinger following the publication of the EPR paper and Einstein's review of 1936 provide information from the time directly following the EPR paper. Detailed analysis of further correspondence and Einstein's contributions

---

[1]This is highly probable. It seems to be the only way to explain the contrast between the usual clarity and precision in Einstein's own writings and the obscure narrative of Podolsky.

to the book "Albert Einstein: Philosopher—Scientist" provide further insight. Fine demonstrates that Einstein's statistical interpretation was not in conflict with the uncertainty relations. Einstein's ensemble interpretation of Schrödinger's $\psi$ function is analyzed in detail. Even if Fine's "prism" interpretation does not correspond to the specifications given in Chap. 8 the conclusions reached are basically the same. In particular, Fine argues convincingly that the deterministic hidden variables theories affected by Bell's theorem have no relation to Einstein's intentions. Einstein's understanding was fully consistent with the physical content of the fundamental laws of Quantum Theory derived by Born and Jordan. But in contrast to Born, Einstein wanted to go much further towards a unified field theory, which should contain Quantum Theory as limiting case only.

Although Podolsky's twisted arguments presented in the EPR paper did not expose Einstein's intentions concerning the inherent statistical nature of Quantum Theory clearly, the attack on the Copenhagen doctrine of unavoidable disturbances caused by measurements was recognized by Bohr. Fine's analysis of Bohr's reaction and the resulting Bohr-Einstein controversy again takes the defense of Einstein. Fine exposes the predominant opinion - that Bohr convincingly refuted the conclusions of the EPR paper—to be a myth. Fine: "I think it is fair to conclude that the EPR paper did succeed in neutralizing Bohr's doctrine of disturbance. It forced Bohr to retreat to a merely semantic disturbance and thereby it removed an otherwise plausible and intuitive physical basis for Bohr's ideas." Einstein remained consistent in his sharp criticism of the Copenhagen narrative; as an example Fine cites Einstein's letter of 5 July 1952 to D. Lipkin: "This theory reminds me a little of the system of delusions of an exceedingly intelligent paranoiac, concocted of incoherent elements of thoughts".

The growing unease among historians and philosophers of Science about Bohr's oracle-like statements is expressed by **Don Howard** in his contribution to **"Niels Bohr and Contemporary Philosophy"** (ed J. Faye and H J. Folse; Dordrecht, Kluwer 1994). The introduction to the article "What makes a Classical Concept Classical?" starts with:

"There was a time, not so long ago, when Niels Bohr's influence and stature as a philosopher of physics rivaled his standing as a physicist. But now there are signs of a growing despair—much in evidence during the 1985 Bohr centennial—about our ever being able to make good sense out of his philosophical views." Howard criticizes "Bohr's self-appointed spokespeople" for having read into Bohr's statements their own views instead of analyzing carefully what Bohr intended to say. He then sets himself the task to perform such a careful and unprejudiced analysis in order to "reconstruct from Bohr's words a coherent philosophy of physics". A difficult task indeed, and varying conclusions reached in previous attempts by others are cited. Howard concentrates his efforts on Bohr's doctrine of classical concepts. It is doubtful that Bohr himself would have appreciated, in particular since, after having analyzed Bohr's ideas, Howard poses the question "Is the doctrine of classical concepts correct?". His answer is: "My opinion is that it is not".

Bohr's statements were not only difficult to understand, they were not meant to be understood in any concrete way, such that an answer to the question Howard asked—'correct ot not?'—could be given; ambiguity is an essential part of the oracle.

Similar conclusions are reached by **Mara Beller** in her book **"Quantum Dialogue, The Making of a Revolution"** (The University of Chicago Press, 1999). The book primarily deals with the struggle for the interpretation of the physical content of matrix and wave mechanics after their appearance in 1925/26. Part I deals with the emergence of the Copenhagen interpretation between 1925 and 1927; Part II with its rhetorical consolidation during the following decades.

The portrayal of the development of Quantum Theory up to 1925/26 is essentially similar to the orthodox accounts by Jammer, van der Waerden, and Mehra-Rechenberg. Beller describes Quantum Mechanics to be unrelated to quantization of the radiation field. She assumes matrix mechanics to originate from the concepts of the old quantum theory, leading directly to Heisenberg's reinterpretation paper of July 1925. Heisenberg is portrayed as the originator of matrix mechanics. Beller's conclusions about the consistency of the Copenhagen interpretation, however, differ strongly from those of the orthodox accounts. In sharp contrast, Beller views the Copenhagen interpretation to be "philosophically deeply unsatisfactory".

Although large part of Beller's criticism coincides with the assessments of the present book, essential differences are numerous. An example: Beller's understanding of the fundamental laws derived by Born and Jordan in September 1925 (BJ-2) is incompatible with that of their originators. Originally, according to Beller: "Matrix mechanics was viewed as a discrete deterministic theory". Beller's error—equating conservation laws with determinacy—is repeated in her discussion of the uncertainty relation, as described below.

Beller's final understanding of the uncertainty relations is equivalent to Heisenberg's: She agrees with Heisenberg's assessment that Quantum Theory is **not** inherently statistical; the uncertainties are blamed on the impossibility to know the initial conditions exactly. But Beller's path to reach this erroneous conclusion is different from Heisenberg's, who had argued that the scattering of a photon by a free particle should necessarily lead to uncontrollable energy and momentum transfers. Beller severely criticizes Heisenberg for this argument[2]; she incorrectly argues that Compton scattering of point-like electrons and photons is deterministic: "There is no way to transcend the classical deterministic framework once it is assumed that photons and electrons are point particles obeying conservation laws". As far as this point is concerned, the fundamental mistake is Beller's, not Heisenberg's. Momentum and energy conservation in Compton scattering still leave a wide range of possible momentum-energy transfers to chance. Within the constraints set by conservation laws, the energy and momentum **transfers** are statistical, as described in Chap. 7 and the appendix.

---

[2]"That a physicist of Heisenberg's stature would make such a mistake is odd enough, but his refusal to correct the mistake, despite powerful criticism from Bohr, becomes incomprehensible...".

Heisenberg's error is more subtle than Beller's. His argument, that **in general** momentum and energy transfers of the measuring process **necessarily** introduce disturbances in the system to be measured, is invalid. As shown in Chap. 7 and the appendix, purely elastic transitions, without any disturbances, are not only possible, but are necessary requirements for the occurrence of diffraction peaks. Quantum Theory is inherently statistical; the result of the individual scattering process is statistical according to the fundamental laws of Quantum Theory, even if the initial conditions are assumed to be known exactly.

Part II of Beller's book contains her strong criticism of the general polyphony of assertions, turnarounds, declarations, contradictions, and changes of mind following the establishment of the Copenhagen interpretation. Criticism is apparent already in the various subtitles, such as: "The Copenhagen Dogma", "The Rhetoric of Finality and Inevitability", and "The Myth of Wave-Particle Duality". Beller points out that Bohr uses complementarity without precisely defined meaning; "what was suggestive and vague became, merely by virtue of repetition, rigorous and compelling". Ambiguity, "enveloped in a fog of profundity", is advocated by Bohr to be a virtue. Bohr's statements—even when they were unintelligible—were taken to represent Bohr's deeper wisdom and understanding. The section on "Bohr and Hero Worship" contains:"The legend that Bohr had some access to nature's secrets, quantitatively different from that of other mortals, directly discouraged critical dialog".

If much of Beller's assessment in part II coincides with the conclusions reached in the present book, an important disagreement remains. Beller portrays Göttingen joining Copenhagen to form a united front enforcing the Copenhagen interpretation. As described before, Born-Jordan (= Göttingen) and Bohr-Heisenberg (= Copenhagen) differed essentially in physical understanding and interpretation of the new theory. The present book directs its criticism towards the Copenhagen additions and alterations of the Göttingen (= Born-Jordan) theory.

**Alexeij Kojevnikov's** contribution "**Philosophical Rhetoric in Early Quantum Mechanics 1925–1927**" (in "Weimar Culture and Quantum Mechanics", ed. A. Kojevnikov, C. Carson, H. Tischler; London and Singapore, 2011) is closer, although he, too, seems to be unaware of the path pursued by Born and Jordan before July 1925. The elimination of the space-time continuum and—as a consequence—indeterminacy of quantum transitions had been Born's program well before 1925; yet Kojevnikov associates Born's advocacy of indeterminacy only with his scattering paper of July 1926. But Kojevnikov recognizes the different understanding of Born-Jordan compared to that of Bohr-Heisenberg, who viewed matrix elements and probabilities as part of a formalism rather than a fundamental principle of the new theory. Kojevnikov:"On their basis of the shared'formalism' of quantum mechanics, its major spokesmen advanced de facto diverging interpretational claims".

Kojevnikov confirms the growing skepticism towards Bohr's philosophy: "Contemporaries overwhelmingly perceived Bohr as the ultimate winner in the interpretational debate over the opposition from Einstein and Schrödinger. Philosophers, who analyze the dispute today, find it hard to explain from a logical point of view".

# General Conclusions

When Max Planck gave the start to the "quantum age", he did so with great hesitation; the introduction of the new fundamental constant, the quantum of action, was made out of mathematical necessity to reproduce the functional form of the "Normalspektrum", replacing the arbitrary assumption of his previous phenomenological derivation. But Planck's mindset concerning natural phenomena was firmly rooted in classical thinking; he hoped that quantization could eventually be replaced by some classical explanation, requiring only minor adjustments about the interaction of radiation with matter.

Einstein's proposal to decompose electromagnetic radiation into quantized particles, absorbed and emitted as indivisible entities, constituted a revolution, which—for Planck and the vast majority of the scientific community—went too far. Wave behavior of radiation seemed to be firmly established and irreconcilable with Einstein's light quanta. The skepticism towards quantized radiation extended even beyond 1925, when the final breakthrough was achieved by Born, Heisenberg, and Jordan. The application of the new quantum laws to quantization of the electromagnetic field by Born and Jordan (BJ-2) at first was "either totally ignored or viewed as a slight attack of craziness" ("leichter Anflug von Verrücktheit", Born, My Life 1968). When Schrödinger proposed wave mechanics in 1926, not only Schrödinger himself, but also Planck and many others belonging to the old guard, hoped that wave mechanics could provide a way back to classical physics, eliminating quantization altogether. Planck finally conceded that: "A new scientific truth usually does not gain general acceptance, because its opponents finally declare themselves to be convinced; it is rather that the opponents gradually die out and the new generation is acquainted with the new truth from the start" (Max Planck, wissenschaftliche Selbstbiographie, 1948).

The old quantum theory started from Planck's point of view, retaining classical pictures about mechanics and electrodynamics. The mechanical behavior of electrons alone was explored; the coupling to the radiation field was left to a postulate, Bohr's "frequency condition". Bohr tried to guess the intra-atomic dynamics, and, quite naturally, directed his attention towards the simplest possible system, the hydrogen

© The Author(s) 2017
H. Capellmann, *The Development of Elementary Quantum Theory*,
SpringerBriefs in History of Science and Technology,
DOI 10.1007/978-3-319-61884-5

atom. The electron circling the nucleus was a natural choice and Bohr's heuristic quantum conditions were able to reproduce the spectroscopic results. What seemed to be a success at first, again quite naturally, was taken as encouragement to pursue this path further, and more and more assumptions were added to reproduce the spectroscopic data of other elements. Although this path did not lead to the desired breakthrough, the twelve years between 1913 and 1925 were probably necessary to prepare the scientific community for the radical changes ahead. The classical concepts of continuity and determinism, which formed the basis for classical equations of motion, had failed, showing the necessity to look for more radical ideas.

It was the failure of the old quantum theory which induced Born to abandon the space-time continuum and to set up the radical "program", laid out in his book "Vorlesungen über Atommechanik" in 1924: Rejecting all speculations about unobservable intra-atomic dynamics, the "true quantum laws" should contain relations between observable quantities; classical laws should apply to macroscopic averages only; quantum dynamics was to be described by discontinuous and statistical laws; quantum states were to be characterized by quantum numbers; precise values of physical variables should no longer be associated with the elementary constituents of the quantum world!

This "program" was conceived before the mathematical formulation of the theory; it was inspired by what had been directly observable about the quantum world. Almost all information available resulted from radiative transitions. The detailed analysis of experimental results about the interaction of radiation with matter had led Einstein to postulate that radiation itself is quantized, consisting of elementary objects having particle character. Einstein had confirmed this hypothesis, showing that thermal equilibrium between radiation and matter required the existence of light quanta, characterized by their energies and momenta. Bose showed that Planck's radiation law was fully consistent with the particle concept of light quanta. When Born made the first step towards the "true quantum laws", Einstein's physical concepts defined the direction to follow; the paper "Quantum Theory of Radiation (Zur Quantentheorie der Strahlung)" of 1916/17 provided the starting point. But whereas Einstein had refused to take the final step, Born accepted discontinuities and purely statistical behavior as fundamental principles.

The new Quantum Theory developed by Born, Heisenberg and Jordan relied on quantization of the action variable; the entire theory was constructed from this principle. Quantization of the action variable required discontinuous dynamics; discontinuities implied statistical behavior. The new quantum laws contained in commutation relations and quantum theoretical equations of motion were built from these principles. Born introduced "Quantum Mechanics"; Born and Jordan recognized that quantization had to apply to all of physics, "Quantum Mechanics" and "Quantum Optics"; they attributed the "basic laws of Quantum Optics" to Einstein. The new quantum laws were applicable to all physical processes, "Quantum Mechanics" and "Quantum Optics" and to their mutual coupling.

The initial mathematical formulation of the new quantum theory was derived from the preceding physical understanding: Nature is discontinuous and statistical; as was the resulting theory, matrix mechanics. Continuous representations followed quickly, generating great flexibility in mathematical methods, which could be adapted to the specific problems to be treated. Unfortunately, this was not only used as mathematical advantage, it also led to misunderstandings about the physical content. Schrödinger's reaction, "I felt deterred, if not to say repelled, by the apparently very difficult methods of transcendental algebra and the lack of illustrative clarity", reflects the attitude of many contemporaries towards the new mode of thought. Discontinuous and statistical behavior was totally opposite to what Schrödinger called "Das räumlich zeitliche Denken" (the mode of thought based on continuity in space and time), and Schrödinger concluded: "We are not really able to change the modes of thought and if we cannot understand within these modes of thought, then we cannot understand at all." Schrödinger's waves were welcomed by Bohr to propose "complementarity": Electrons sometimes should behave as particles and sometimes behave as waves. Complementarity became the principle obstacle to understanding Quantum Theory; instead of explaining any basic principle, "complementarity" covered quantum physics with a veil of mysticism.

When progress in experimental techniques provided access to the quantum world, this was new territory, and the old concepts failed to describe the new findings. Born accepted that old prejudices had to be abandoned; the understanding of the quantum world does require to change the traditional mode of thought in regard to natural phenomena. And that is what Born did, when he abandoned the space-time continuum. The quantum laws derived by Born and Jordan have stood the test of time; the experimental evidence collected over the past 100 years have confirmed their basic principles: Nature on the elementary level is discontinuous and statistical; classical laws are approximately valid for macroscopic averages.

**Acknowledgements** Many thanks to Efim Kats for helpful comments.

# Appendix

## Scattering Processes and the Basic Quantum Laws

Scattering experiments are instructive to demonstrate the basic principles of quantum physics; the relation between the basic laws and elementary scattering processes is particularly simple and direct. The use of the energy-momentum representation will demonstrate the independence of physical content of particular mathematical representation. So-called interference will be shown to be a mathematical artifact of Schrödinger's position representation. Actual measurements of quantum uncertainties of particle position will be discussed to demonstrate that Heisenberg's disturbance-related explanation is erroneous.

The Hamiltonian

$$H = H_0(\hat{\mathbf{p}}) + H_0(X) + V(X, \hat{\mathbf{r}}) \tag{A.1}$$

describes the coupled system of a particle with some other system $X$. $H_0(\hat{\mathbf{p}})$ is the Hamiltonian of the free particle, $H_0(X)$ the Hamiltonian for the system X. The coupling is taken to be a scalar field $V(X, \hat{\mathbf{r}})$. $\hat{\mathbf{p}}$ and $\hat{\mathbf{r}}$ are particle momentum and position operators with commutation relation

$$[\hat{\mathbf{p}}\hat{\mathbf{r}} - \hat{\mathbf{r}}\hat{\mathbf{p}}] = \frac{\hbar}{i}\mathbf{1}, \tag{A.2}$$

as introduced by Born and Wiener; e. g. $\hat{\mathbf{p}}$ is an operator with eigenvalues $\mathbf{p}$ and eigenvectors $|\mathbf{p}\rangle$; explicitly: $\hat{\mathbf{p}}|\mathbf{p}'\rangle = \mathbf{p}'|\mathbf{p}'\rangle$.

We address the experimental setup, where a source provides particles of initial momentum $\mathbf{p}_i$ to be scattered off a target system described by initial quantum number $X_i$. The scattering process produces final states of momentum $\mathbf{p}_f$, and target quantum number $X_f$. The momenta are the measured quantities. Take initial and final states as eigenstates of $H_0(\hat{\mathbf{p}}) + H_0(X)$. The "equation of motion" for $\hat{\mathbf{p}}$

© The Author(s) 2017
H. Capellmann, *The Development of Elementary Quantum Theory*,
SpringerBriefs in History of Science and Technology,
DOI 10.1007/978-3-319-61884-5

$$\langle \mathbf{p}_f, X_f \mid \hat{\dot{\mathbf{p}}} \mid X_i, \mathbf{p}_i \rangle = \frac{i}{\hbar} \langle \mathbf{p}_f, X_f \mid [H\hat{\mathbf{p}} - \hat{\mathbf{p}}H] \mid X_i, \mathbf{p}_i \rangle \qquad \text{(A.3)}$$

reduces to

$$\langle \mathbf{p}_f, X_f \mid \hat{\dot{\mathbf{p}}} \mid X, \mathbf{p} \rangle = (\mathbf{p}_i - \mathbf{p}_f) \frac{i}{\hbar} \langle \mathbf{p}_f, X_f \mid V(X, \hat{\mathbf{r}}) \mid X_i, \mathbf{p}_i \rangle . \qquad \text{(A.4)}$$

Compare the classical and quantum versions of the equation of motion: In classical physics $\dot{\mathbf{p}}$ is defined by an infinitesimally small momentum interval $d\mathbf{p}$ divided by an infinitesimally small time interval $dt$ . Quantization of the action variable requires all physical variables to change discontinuously; the quantum equation of motion specifies the admissible transitions. For the example chosen here, the allowed momentum intervals $\Delta \mathbf{p} = (\mathbf{p}_f - \mathbf{p}_i)$ are determined by the factor $i/\hbar \langle \mathbf{p}_f, X_f | V(X, \hat{\mathbf{r}})|X_i, \mathbf{p}_i \rangle$, which has dimension $(dt)^{-1}$; the corresponding transition probability per unit time is proportional to $|\langle \mathbf{p}_f, X_f | V(X, \hat{\mathbf{r}})|X_i, \mathbf{p}_i \rangle|^2$.

Initial and final states are product states (e.g. $\mid X_i, \mathbf{p}_i \rangle \ = \mid X_i \rangle \mid \mathbf{p}_i \rangle$); we define

$$V_{X_f, X_i}(\hat{\mathbf{r}}) = \langle X_f \mid V(X, \hat{\mathbf{r}}) \mid X_i \rangle. \qquad \text{(A.5)}$$

The evaluation of the remaining matrix element $\langle \mathbf{p}_f \mid V_{X_f, X_i}(\hat{\mathbf{r}}) \mid \mathbf{p}_i \rangle$ proceeds by Fourier expansion

$$\langle \mathbf{p}_f \mid V_{X_f, X_i}(\hat{\mathbf{r}}) \mid \mathbf{p}_i \rangle = \langle \mathbf{p}_f \mid \int_{\mathbf{q}} \tilde{V}_{X_f, X_i}(\mathbf{q})\, e^{-i\mathbf{q} \cdot \hat{\mathbf{r}}} \mid \mathbf{p}_i \rangle. \qquad \text{(A.6)}$$

In momentum representation ($\hat{\mathbf{r}} = -\frac{\hbar}{i} \mathbf{V}_\mathbf{p}$) the Taylor expansion of $|\mathbf{p} + \hbar\mathbf{q}\rangle$ may be written as $e^{-i\mathbf{q} \cdot \hat{\mathbf{r}}} |\mathbf{p}\rangle$. We obtain

$$\langle \mathbf{p}_i + \hbar\mathbf{q}, X_f \mid \dot{\mathbf{p}} \mid X_i, \mathbf{p}_i \rangle = i\, \mathbf{q}\, \tilde{V}_{X_f, X_i}(\mathbf{q}). \qquad \text{(A.7)}$$

The transition probability for the particle to be scattered with momentum transfer of $\hbar\mathbf{q}$ and the scattering system making a transition from $X_i$ to $X_f$ is proportional to $\mathbf{q}^2 |\tilde{V}_{X_f, X_i}(\mathbf{q})|^2$. Although this result may be obtained using any representation, the momentum representation is best suited to make the connection between physical content and mathematical formalism clearest. Initial and final particle momenta are the observable quantities, their representation by real variables avoids the misconception, that wave functions might be more than mathematical tools.

## Particle Properties and Diffraction Phenomena

The occurrence of diffraction peaks, e. g. in Bragg scattering of photons or other particles, has often been argued to necessarily require wave properties. Bohr's rejection of Einstein's light quanta was based on this argument; similarly Bohr argued that the observation of diffraction in electron scattering required a wave property of electrons. This conclusion led Bohr to postulate dual character of electrons, particle—and wave-like (and similarly for other particles). We shall show that these conclusions are erroneous.

We specify the target system to be a crystal. In order to establish the direct relation between quantum equation of motion and the observed scattering processes, we discuss the case of weak interaction between particle and crystal; under these conditions (typically satisfied in neutron scattering) the scattering process consists of a single elementary quantum transition and is directly described by the quantum equation of motion.[1] The experimental setup is the following:

A source provides particles with energy $E_i$ and momentum $\mathbf{p}_i$ directed towards the crystal; energy and momentum of the scattered particles are recorded. At resolution sufficient to observe single particle scattering the principle observations are: Due to weak interaction most particles are not scattered at all. The particles which are scattered are distributed statistically over a wide range of directions; certain special directions ("Bragg peaks") are statistically preferred, however, and may be distinguished from the background contributions. The energy $E_f$ and the absolute value $|\mathbf{p}_f|$ of the scattered particles contributing to Bragg peaks are unchanged, resulting from purely elastic scattering. Experimental demonstrations are available in form of two short videos produced by the Juelich Center for Neutron Science.[2]

The theoretical conclusions are directly obtained from the results derived above. According to the quantum equation of motion, the total scattering probability $W_{tot}(X_i)$ for given initial target quantum number $X_i$ is proportional to the sum over all possible final states $|X_f\rangle$. Of special importance are scattering processes, which do not induce any change in crystal quantum numbers; initial and final state of the crystal are identical, $(X_f = X_i)$,

$$W_{tot}(X_i) \sim \int_{\mathbf{q}} \mathbf{q}^2 \left( \mid \tilde{V}_{X_i,X_i}(\mathbf{q}) \mid^2 + \sum_{X_f \neq X_i} \mid \tilde{V}_{X_f,X_i}(\mathbf{q}) \mid^2 \right). \qquad (A.8)$$

The essential condition for Bragg scattering contributions: Only those events, which "do not leave any trace in the crystal" (i. e. the contributions from $X_f = X_i$) may contribute to Bragg scattering, whereas all other scattering (i.e. the sum over $X_f \neq X_i$) contributes to the rather structureless background.[3]

---

[1] The qualitative conclusion below will also be valid for the scattering of photons and other particles.

[2] The first video demonstrates the basic principles of Bragg scattering (www.fz-juelich.de/SharedDocs/Videos/PORTAL/EN/JCNS-Neutron-Scattering-Part-1.html). Background contributions due to spin dependent couplings between neutron and nuclear spins are demonstrated in the second video (www.fz-juelich.de/SharedDocs/Videos/PORTAL/EN/JCNS-Neutron-Scattering-Part-2.html). The videos may be downloaded for lectures and self-study from www.fz-juelich.de/jcns/EN/Leistungen/Education/Videos/_node.html

[3] An example for background contributions in neutron scattering: The vectorial interaction between neutron spin and nuclear spins provides finite probabilities for nuclear spin flips. The total scattering probability for nuclear spin transitions is a sum over the scattering probabilities from individual

Purely elastic scattering events require energy conservation; $\mathbf{p}_i^2 = \mathbf{p}_f^2 = (\mathbf{p}_i + \hbar\mathbf{q})^2$. Assume the crystal to be in a state of perfect crystalline periodicity, characterized by a set of real space vectors $\mathbf{L}$, such that for integer $n$

$$V_{X_i,X_i}(\mathbf{r} + n\mathbf{L}) = V_{X_i,X_i}(\mathbf{r}). \tag{A.9}$$

Nonvanishing Fourier coefficients $\tilde{V}_{X_i,X_i}(\mathbf{q})$ will be restricted to $\mathbf{q} = \mathbf{Q}$, where

$$\mathbf{Q} \cdot \mathbf{L} = n\, 2\pi, \tag{A.10}$$

$n$ integer. Discrete translational symmetry selects special momentum transfers $\hbar\mathbf{Q}$, which, combined with energy conservation $\mathbf{p}_i^2 = \mathbf{p}_f^2 = (\mathbf{p}_i + \hbar\mathbf{Q})^2$, characterize the Bragg peaks.[4] Whereas full translational symmetry would require momentum to be conserved, discrete translational symmetry conserves "**quasi momentum**"; a particle of momentum $\mathbf{p}$ may be scattered into $\mathbf{p} + \hbar\mathbf{Q}$; the scattering probabilities are determined by $|\tilde{V}_{X_i,X_i}(\mathbf{Q})|^2$.

"Quasi momentum conservation" is the direct consequence of Born's quantization condition: The products of the symmetry vectors $\mathbf{L}$ and the allowed momentum transfers $\hbar\mathbf{Q}$ are equal to the change in action variables, which, according to Born's quantization condition, have to be integer multiples of Planck's constant; $\hbar\mathbf{Q} \cdot \mathbf{L} = nh$.

The results above were obtained using the momentum representation; identical results may, of course, be obtained using any other representation, for example Schrödinger's position representation and wavefunctions. Since there are infinitely many canonical transformations leaving the commutation relations invariant, there are infinitely many possible mathematical representations. We may choose any one of them according to mathematical convenience or other criteria; but no logical argument can be advanced to justify the conclusion: "Diffraction phenomena may be described mathematically using wavefunctions and thereby prove wave properties of particles". Similarly, there is no logical necessity to invoke concepts like "self-interference". An advantage of the momentum representation, which represents the **observable** quantities by real numbers, might be that its use prevents such erroneous conclusions.

Bohr's Complementarity principle—advocating dual properties, wave and particle like—originated from Bohr's belief that diffraction required interference of waves. The second argument, central to the Copenhagen interpretation, led to the so-called "measurement problem": Measuring processes—according to Heisenberg and Bohr—necessary cause disturbances in the system to be measured. This argument, too, is erroneous. As was shown above, the essential condition for all scattering

---

(Footnote 3 continued)
lattice sites. The same argument applies to all processes, which cause localized transitions in the crystal.

[4] Real samples contain crystalline disorder, which leads to finite widths of Bragg peaks and additional background contributions.

processes contributing to diffraction peaks is purely elastic scattering, no change in quantum numbers of the crystal may occur. The very existence of diffraction phenomena already disproves the claim made by Heisenberg and Bohr. The Copenhagen interpretation—based on disturbance induced uncertainties and dual-properties—contains internal inconsistencies. Scattering causing disturbances cannot contribute to diffraction; observed diffraction does not require wave-like properties.

## Measuring the Quantum Mechanical Position Uncertainty

Quantum uncertainties are not due to disturbances introduced in measuring processes, they are constitutive elements of quantum physics, as was originally stated by Born and Jordan. Purely elastic (i.e. disturbance free) scattering processes may actually be used to **measure** the quantum uncertainties of positon. We present a simple example, which has become standard practice in neutron scattering, not only measuring the average positions of crystal nuclei but also their quantum uncertainties.

The nuclear interaction between the neutron and the crystal nuclei[5] may be considered to be a contact interaction and written as a sum over lattice sites $l$

$$V(\hat{\mathbf{R}}_l, \hat{\mathbf{r}}) = \sum_l b_l \, \delta(\hat{\mathbf{r}} - \hat{\mathbf{R}}_l). \tag{A.11}$$

The $\hat{\mathbf{r}}$ is the neutron position operator, the $\hat{\mathbf{R}}_l$ are position operators of crystal nuclei, the $b_l$ are the scalar coupling constants. Let the crystal be in a state corresponding to quantum numbers $X_i$ and average positions of nuclei $\mathbf{R}_l^i$. The diagonal matrix element of the interaction $V(\hat{\mathbf{R}}_l, \hat{\mathbf{r}})$ over $|X_i\rangle$ will result in

$$V_{X_i, X_i}(\hat{\mathbf{r}}) = \sum_l b_l \, f_i(\hat{\mathbf{r}} - \mathbf{R}_l^i), \tag{A.12}$$

where the function $f_i(\mathbf{r} - \mathbf{R}_l^i)$ represents the position uncertainty of the nucleus at lattice site $l$ in the state $|X_i\rangle$. The further evaluation may proceed as in the preceding section on Bragg scattering. The purely elastic scattering probability for momentum transfer $\hbar\mathbf{q}$ is

$$| \tilde{V}_{X_i, X_i}(\mathbf{q}) |^2 = | \sum_l e^{i\mathbf{q}\cdot\mathbf{R}_l^i} b_l \cdot \tilde{f}_{i,l}(\mathbf{q}) |^2, \tag{A.13}$$

where the function $\tilde{f}_{i,l}(\mathbf{q})$ is the Fourier transform of $f_{i,l}(\mathbf{r})$, and $|f_{i,l}(\mathbf{r})|^2$ gives the probability distribution for the nuclear position at site $l$ in the state $|X_i\rangle$. For temperature $T$ tending towards zero and the crystal being in its ground state ($i = 0$), $f_{0,l}(\mathbf{r})$ represents the nuclear position uncertainty due to zero point uncertainties.

---

[5]Only the scalar part of the interaction is considered; the spin dependent vectorial interaction does not contribute to Bragg scattering and plays no role for the arguments below.

For perfect crystalline periodicity the functions $\tilde{f}_{0,l}(\mathbf{q})$ are identical for equivalent lattice sites; the sum over $l$ on the right hand side of the equation above will guarantee that $\tilde{V}_{X_0,X_0}(\mathbf{q})$ vanishes except for the special values $\mathbf{q} = \mathbf{Q}$ (the Bragg peaks). These $\delta$-peaks will attain finite widths in real crystals due to finite grain size of crystallites and crystalline disorder; furthermore, the accuracy of neutron momenta is restricted by experimental resolution and quantum uncertainties. To lowest order, the total intensity of the various Bragg peaks will be unaffected. Their intensities provide a finite number of Fourier components of the functions $f_{0,l}(\mathbf{r})$, and a large enough number of Bragg peaks measured enables a reasonable reconstruction of $f_{0,l}(\mathbf{r})$. At $T = 0$ the position uncertainty is due to zero point quantum uncertainties. For oscillators at $T = 0$ the minimum value of the mean square deviations allowed by the fundamental laws is reached; for finite temperature, states of higher energy are excited, the experiment measures the thermal average and the position uncertainty increases.

A final remark concerning the use of momentum-energy eigenvectors in the calculations above may be helpful. As mentioned already repetedly, physical systems—in contrast to mathematical models—never possess exactly sharp physical variables due to necessary quantum uncertainties. Furthermore, experimental uncertainties due to finite resolution are typically even much larger than the minimal quantum uncertainties required by the commutation relations. Nevertheless, momentum-energy uncertainties of the incoming neutrons do not invalidate the conclusions reached above. Born's quantization condition—changes in action variables are restricted to integer values of $h$—requires the **momentum transfers** to be independent of the initial momentum uncertainties. The equivalent argument holds for the **energy transfers**, as will be discussed below. Neither quantum nor additional experimental uncertainties of the incoming neutrons prevent the possibility to measure the quantum uncertainty of nuclear positions. The experiment measures the energy and momentum transfers, and the conclusions reached depend on these transfers, not on perfectly defined initial values.

## Remarks Concerning Field Quantization

The momentum-energy representation may be used to obtain a **"shortcut" to field quantization**. Take the classical interaction between a particle and the field to be of simple scalar form $V(\mathbf{r}, t)$, where $\mathbf{r}$ is the position variable of the particle. Although electric and magnetic fields are vector fields and the coupling is not a simple scalar coupling, this is not important for the following; the arguments below may be applied to quantization of all types of classical fields.

Take the Fourier expansion of the classical field $V(\mathbf{r}, t)$:

$$V(\mathbf{r}, t) = \int_{\mathbf{q},\omega} \tilde{V}(\mathbf{q}, \omega) e^{-i(\mathbf{qr}-\omega t)}. \tag{A.14}$$

The transition to quantum theory replaces the classical variables by operators; using the "momentum-energy representation", $\mathbf{r}$ is replaced by the operator $\hat{\mathbf{r}} = -\hbar/i \; \nabla_{\mathbf{p}}$ and $t$ is replaced by $\hat{t} = \hbar/i \cdot d/dE$. The perturbation $V(\hat{\mathbf{r}}, \hat{t})$ acting on a particle state $|E, \mathbf{p}\rangle$ of well defined energy $E$ and momentum $\mathbf{p}$ results in

$$V(\hat{\mathbf{r}}, \hat{t}) \; |E, \mathbf{p}\rangle = \int_{\mathbf{q}, \omega} \tilde{V}(\mathbf{q}, \omega) \; |E + \hbar\omega, \mathbf{p} + \hbar\mathbf{q}\rangle. \tag{A.15}$$

A field of frequency $\nu$ and wavelength $\lambda$ may cause the particle to make a quantum transition with energy transfer $\Delta E = \hbar\omega = h\nu$ and momentum transfer $|\Delta\mathbf{p}| = \hbar|\mathbf{q}| = h/\lambda$; the transition probability is proportional to $|\tilde{V}(\mathbf{q}, \omega)|^2$. $\Delta E/\nu$ and $\lambda \cdot |\Delta\mathbf{p}|$ represent the change in action variable, which—as required—is equal to $h$.

Quantization of action has to affect **all** physical variables, i. e. must have consequences on the field as well: Born's quantization condition applied to a field of frequency $\nu$ and wavelength $\lambda$ requires the existence of quanta with energy $\epsilon = h\nu$ and momentum $p = h/\lambda$, respecting energy and momentum conservation for the individual elementary quantum transition. This reflects Einstein's reasoning, when he postulated photons in 1905: The experimental observations of the interaction between radiation and matter, in particular the energy and momentum exchange between the field and point like particles, require the existence of radiation quanta having particle character.

# Author Index

© The Author(s) 2017
H. Capellmann, *The Development of Elementary Quantum Theory*,
SpringerBriefs in History of Science and Technology,
DOI 10.1007/978-3-319-61884-5

# Subject Index

© The Author(s) 2017
H. Capellmann, *The Development of Elementary Quantum Theory*,
SpringerBriefs in History of Science and Technology,
DOI 10.1007/978-3-319-61884-5

Printed in the United States
By Bookmasters